CAVE EXPLORING

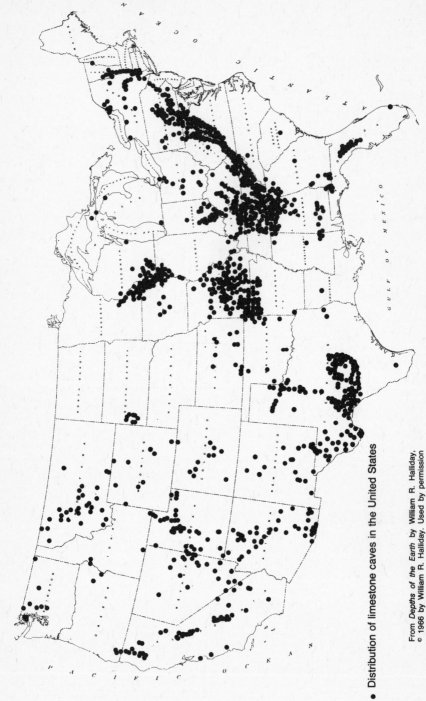

• Distribution of limestone caves in the United States

From *Depths of the Earth* by William R. Halliday, © 1966 by William R. Halliday. Used by permission of Harper & Row.

CAVE EXPLORING

Jennifer Anderson

*Foreword by John A. Stellmack, Past President
National Speleological Society*

ASSOCIATION PRESS / New York

CAVE EXPLORING

Copyright © 1974 by Jennifer Anderson

Association Press, 291 Broadway, New York, N.Y. 10007

International Standard Book Number: 0-8096-1889-3
Library of Congress Catalog Card Number: 74-8638

Library of Congress Cataloging in Publication Data
Anderson, Jennifer, 1942 –
Cave exploring.
Bibliography: p.
1. Speleology. I. Title.
GB602.A62   796.5'25   74-8638
ISBN 0-8096-1889-3

Printed in the United States of America

CONTENTS

FOREWORD

If someone had told me many years ago, as I emerged from my first muddy Pennsylvania cave, "Well, now that you know what it's like underground, you'll have to learn how to get around properly down there," my response would have been, "You mean you have to learn techniques to do *that!*" But as I visited more and more caves and encountered a widening variety of challenges, I came to appreciate the fact that there are efficient and safe methods for almost every caving technique, even for crawling on one's stomach through the mud. I also discovered that with the efficiency and safety came unsuspected benefits. Most neophyte cavers treat their first underground experiences with varying degrees of timidity and fear, which is as it should be. However, as the neophyte gains in experience and assurance, and as he becomes more proficient in his caving techniques, he will find these fears slipping away. The physical aspects of moving through the cave and using the needed equipment properly at the correct time will become second nature to him. It is then that he is able to most fully enjoy the hidden sights and wonders which the cave has waiting for him.

Unfortunately, some individuals new to cave exploration find themselves in trouble *before* they learn this. The bane of the experienced cave explorer is the Sunday afternoon "flashlight caver" who gets his name in the paper because his batteries go dead and he must be rescued. Because of these individuals, and remembering poor Floyd Collins (God rest his caver's soul), experienced cave explorers spend considerable time trying to convince the noncaving public (especially mothers and wives) that the sport need not be dangerous.

On the other hand, like any other outdoor physical activity, caving can be dangerous if you don't know what you are doing. When underground, your isolation from the rest of the world is essentially complete and you are entirely dependent upon the knowledge and equipment possessed by you and your group. It is imperative that the prepared cave explorer not only be aware of the possible difficulties he may encounter below the surface, but that he should also have knowledge of the efficient use and limiting factors of both his own body and the equipment he carries—be it his light source, the clothing he wears, or his rope.

It is said that one learns from past mistakes but this is not necessarily true in cave exploration. The "learning" mistake of the surface can too often have dire consequences in the isolation of the cave environment. Rather, one's goals in learning about caving should be to discover: first, how and where mistakes can happen; second, how to avoid them; and third, the best thing to do when they occur (as they probably will). Becoming familiar with proper caving techniques is a first step toward these goals.

Cave exploration can be a very demanding activity and the difficulties encountered underground will constantly challenge both your endurance and your capabilities. But that is an integral part of the pleasure of caving. An obstacle overcome is a victory won, which affords a uniquely pleasurable experience for a human being. It is equally important, however, to be fully aware of your own deficiencies and to know when to say to yourself, "Enough! Turn back!" An over-extended caver, be it that he is suffering from fatigue, cold, or lack of proper equipment or experience, not only jeopardizes his own safety but can also endanger others in his party. In an activity such as caving, where the safety of the group so highly depends upon the good judgment of each individual, every person should be aware of the line at which he stops being an asset to the group and becomes a liability instead. One does *not* play "chicken" underground.

Conservation, unfortunately, is all too often not considered an essential part of caving technique, but any cave explorer worthy of the name knows just how important conservation really is. And this doesn't mean just breaking rock formations or dumping trash in a cave, bad as these things may be. A true caver appreciates, cherishes and protects *everything* within the cave and its immediate environs. What is observed in a cave today, whether it be animal, vegetable or mineral, took thousands of years, often longer, to reach its present state. It can be destroyed in a moment of thoughtlessness. All cavers should help to preserve these sources of our present enjoyment for those who come after us. Possibly the most fragile thing connected with a cave is the relationship between the cave explorer and the cave-owner. This should be treated with utmost care at all times because, once broken, it is exceedingly difficult to put back together again—that is, if it can be done at all.

Through many years of experience in education, I have concluded that the best learning experience is obtained by a combination of a good textbook and a motivated teacher. You now have the textbook in your hands. If you are a beginner, next seek the teacher, an experienced caver. The National Speleological Society (Cave Avenue, Huntsville, Alabama 35810) has active chapters throughout the country. Contact the society to find the chapter nearest you. You may not only find the teacher you seek and an introduction to a unique and satisfying activity, but also quite likely the beginning of friendships that will last you a lifetime—as I did. Good caving!

John A. Stellmack
Past President
National Speleological Society

INTRODUCTION

Adversity lurks in every cave. This somewhat pessimistic view is justified by a startling number of unpleasant incidents of every sort. People get lost in caves. They become trapped by underground floodwater. They lose control when they rappel. They suffer the terror of being hung up on a rope with Jumars that won't grip or with slipping Prusik knots. These are only a few of the horrors that can greet the greenhorn when he goes caving.

An expert caver, in my opinion, is a person who recognizes adversity in all its underground forms. He stays out of trouble not by avoiding challenging situations, but by meeting the challenges with appropriate skills and equipment. If, despite his preparedness, unforeseen problems occur, the expert knows how to extricate himself from his difficulty. The popular picture of the caving expert, on the other hand, features a person who is knowledgeable about all the latest equipment or caves the deepest pits, or the longest and wettest caves. He is often featured with a carload of caving gadgetry in a four-wheel-drive vehicle. As a matter of fact, both of these types of caving expert can admirably co-exist in the same person. However, from the point of view of survival and enjoyment of caving, it is better to assume the attitudes of the first kind of caver before taking on the features of the popular type of expert. It is, therefore, to cavers who are intent on survival that this book is dedicated.

Only those topics which are relevant to caving safety have been included, except for a brief mention of cave and speleothem formation as it relates to cave conservation. The section on cave conservation was included despite its negligible relationship to safety because it is my unwavering conviction that no opportunity should ever be missed to pass on the "Take nothing but pictures, leave nothing but footprints" message to protect our caves. The subject of caver's relationship with the caver-owner was included because cave safety becomes meaningless if such relationships deteriorate to the extent where cave-owners feel compelled to close their caves to cavers.

ACKNOWLEDGMENTS

To the friends and acquaintances who helped me with this book, I give my sincere thanks. In particular, I wish to thank William Mixon, Dr. F. Wefer, and Alan P. Haarr for their critical reviews of the manuscript. I relied on past discussions with Richard Schreiber to help define and present the most important aspects of rappel and Prusik techniques, and on discussions with Dr. Y. Mason to clarify the advice in the chapter on caving injuries. I would also like to thank Ed Lisowski who posed, with remarkable patience, for the various illustrations in this book.

The professional services of S. Jerde were engaged for the illustrations in the chapter on belaying.

1. Introduction to the Cave

Cave Formation

Caves are formed in limestone in two ways. The first of these is the purely solutional activity of ground water in the zone below the water table. In this zone the rock is saturated with water, and the caves formed in it are known as *Phreatic* caves. In the second way, *Vadose* caves are formed by the other agent, surface water. This surface water absorbs small amounts of carbon dioxide from the air which forms a mild solution of carbonic acid. This becomes stronger as it picks up additional amounts of carbon dioxide from the soil. The limestone is very susceptible to this mild acid which percolates down joint cracks and dissolves out the less resistant beds of rock. Sand and other debris which is carried in the water, acts as an abrasive and hastens the process of cave formation. This surface water activity produces passages with a characteristic shape (see Figure 1). The shape of passages in Phreatic caves is also quite characteristic (see Figure 2). However, the distinction between Vadose and Phreatic passages is often difficult to make in areas where the Phreatic passage has been modified by Vadose activity or where the limestone has been subjected to folding and faulting prior to solution. Caves in a folded, faulted area are generally complex, with passages going off, over, and under other passages. This type of cave often presents more challenging caving than the caves formed in flat-lying limestone which more commonly houses caves of an uncomplicated single-level type.

The ground water dissolves out calcium carbonate from the limestone as it travels downward. When it reaches the open cave atmosphere, some carbon dioxide is given off, and pure calcite is precipitated on the walls, ceiling, and floor of the cave to form the speleothems which add such beauty to many caves. The rate at which these grow depends upon the amount of surface rainfall as well as many other factors. Speleothems can be the product of a few months or many hundreds of years. The caves themselves take thousands of years to form and

15

can be expected to last many thousands of years more if all people who enter caves treat them with the respect they deserve as works of nature.

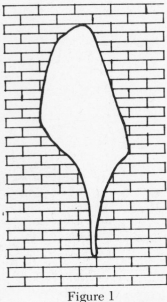

Figure 1
Cross section of Vadose cave.

Little short of dynamite can affect the cave proper, but the speleothems are extremely vulnerable to people, and damage to the speleothems is irreparable. Nearly all damage to speleothems is caused by people. Some of this damage is pure vandalism by sick people whose only reaction to beauty is destruction. Much of the damage, however, is by unthinking cavers, not all of them newcomers, who have never stopped to consider the effect they have on the cave.

Here are a few simple, common-sense, rules of cave behavior and conservation that will ensure the beauty of the caves will linger to be enjoyed by many future generations.

1. Exercise care not to break or soil formations accidentally. Once you have visited a highly decorated cave, you will realize just how easy it is to bump and break delicate straws and stalactites inadvertently, and to trample and break rimstone dams and the beautiful calcite crystal clusters that they frequently contain. A certain amount of this type of damage is inevitable in a well-decorated cave, but it can be minimized. Don't track up one more inch than absolutely necessary. If there is any route around rather than through the formations, take it — even if it is longer and more time consuming. In places where there are delicate formations on a floor that has to be crossed take off your boots before crossing, if possible. This same procedure should also be followed when it is nec-

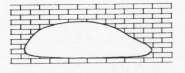

essary to cross very white or delicately colored flowstone. Stockinged feet do much less damage and keep mud off the floor. Avoid using speleothems as handholds.

Figure 2
Cross section of Phreatic cave.

Figure 3
Photograph of bacon rind formations, small stalactites (hanging from bacon rinds) and stalagmites (on the floor) which were found in a West Virginia cave. (Photograph by Diana Northup)

2. Leave nothing in the cave. Spent carbide, spent flashlight batteries, flash bulbs, and candy wrappers are all litter. They are just as unattractive inside a cave as outside a cave, and, in addition, spent carbide is extremely poisonous to cave life. Take the carbide out in a strong plastic bag. Don't leave it outside the cave, either. Farm animals can pick it up when they eat grass and will be poisoned. The spent carbide should be dumped in a regular trash can when you get back to civilization.

Figure 4
Helictites in a West Virginia cave. (Photograph by Diana Northup)

3. Write nothing on walls or formations. Route-finding markers and survey markers should be of the type that can be removed on the way out.

4. All these conservation principles can be practiced right from your first trip. No novice should go underground without the benefit of a short lecture on the dos and don'ts of cave conservation. If a novice or even an experienced caver is seen deliberately breaking formations, he should be escorted out of the cave immediately and expelled from the group. Any further trouble should be reported to the Cave Conservation Committee of the National Speleological Society.

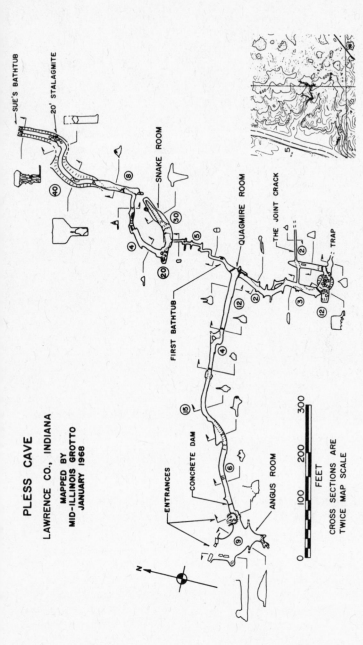

PLESS CAVE

LAWRENCE CO., INDIANA

MAPPED BY
MID-ILLINOIS GROTTO
JANUARY 1968

N

SUE'S BATHTUB
20' STALAGMITE
SNAKE ROOM
THE JOINT CRACK
TRAP
QUAGMIRE ROOM
FIRST BATHTUB
ENTRANCES
CONCRETE DAM
ANGUS ROOM

FEET

0   100   200   300

CROSS SECTIONS ARE
TWICE MAP SCALE

Map of typical, uncomplicated cave with passage cross sections shown at intervals along the passage. Note the section of topographic map in the lower righthand corner with the cave superimposed in black. This gives an indication of the size and location of the cave in relation to surface features.

5. Don't collect cave animals. Populations of cave animals are usually very small and lead a precarious existence. Some species are near extinction. Not only is collecting likely to upset a delicate balance, but it could also deplete the already small populations to the point where they can't recover. Hibernating bats should never be touched. If they are wakened from their hibernation sleep, so much of their stored body energy is used up in the process that they may well be unable to survive the remaining hibernation period. Also, avoid entering a cave that contains a colony of nursing bats. The female bats frequently drop their young to their death if they are disturbed in such circumstances.

Cave-Owner/Caver Relationships

Courtesy is the key word. Leave everything as you find it: gates shut or open (as you found them), cattle unchased, boards or other barriers replaced at a pit or cave entrance after you come out. Sound out the owner's wishes on everything. He may have preferences about where you park your car. He may want you to ask permission to enter the cave every time you come, or to sign a liability release. He may prefer not to be disturbed. Find out. When in doubt, ask permission to enter the cave and if it is denied, don't enter the cave. We rely on the forbearance, patience, and kindness of the cave-owner for the caves we visit. There are no benefits for him in having cavers on his property and he has every right to close his cave to cavers. It is in our best interests not to give him any excuse for doing so.

2.   Personal Caving Equipment

The temperature of the average cave in the United States is cool or even cold and the terrain is rough, so you know that you will need heavy clothing. In each case, however, you should find out in advance all you can about a cave that you propose to visit.

Clothing

You can begin with combination overalls that fit comfortably over a lightweight wool sweater, which should be next to the skin, and old underpants (or long underwear if the trip is likely to involve sitting or standing still for 30 minutes or more). Heavy wool socks under Vietnam combat boots are ideal for footwear. A belt should be worn to keep the overalls fitting comfortably. For this purpose, a leather belt is fine, but it can be improved upon. A 30-foot length of 1-inch tubular nylon or 5/16-inch Goldline can be wound firmly around the waist and tied with a bowline. This makes a belt with multiple uses. A carabiner (an oblong ring to hold a freely running rope, see Figure 12, page 33) should be clipped into it. With this belt carabiner you can hook yourself onto the ladder if you need to rest or to unsnag the ladder, or if you need two hands to take a picture, and so on. This type of belt is also useful as a belay loop. Most useful of all, it is instantly available should a handline be needed. The woolen sweater is recommended because of the superior insulating properties of wool. In such a sweater, you are well protected against overheating or chilling. Even when wet, wool will still keep you warm for quite some time. This does not mean that you cannot become overheated or cold when wearing wool, it simply means that it takes longer for these discomforts to become manifest than for some other fabrics. A long-sleeved thermal knit shirt is the next best choice for a caving shirt. In most instances it is wise to include a spare sweater if the cave includes a "bathtub," or low spot in the path which is normally filled with water. To keep

Figure 6
The well-dressed cave explorer.

this sweater dry you will need a rigid, waterproof container. A 2-lb.
coffee can with a plastic lid or an ammunition box bought at an Army
surplus store will remain watertight during total immersion. In this
way your wet sweater can be replaced before the discomfort of cold
has set in. For caves where the "bathtubs" are numerous or so long
that you will be immersed for stretches of 15 minutes or more, a
neoprene wet suit is advisable because of the possibility of hypothermia,
or loss of heat in the body core. Such a wet suit, however, represents
a considerable investment; so, until you are ready to commit yourself
to wet caving and decide to buy a wet suit, it will be wise to stay
out of very wet caves.

To finish off the outfit, you will need a hard hat and a waterproof
wristwatch. With the garb outlined above (see Figure 6), you are ready
for all but the very wet caves. This represents an ideal outfit, of course,

Figure 7
Vietnam combat boots.

and substitutions can be made. Don't cut corners with either your
boots or your hard hat, but what you put between can come from
your old clothes supply. In the early stages of your caving career
you will probably find yourself doing easy caves and very short trips.
If this is the case stout leather shoes will suffice. Don't wear either
sneakers or loafers, however. Both have an unhandy disposition for
disappearing without a trace in deep mud, leaving their ex-wearer
to get out of the cave in his bare feet as best he can. Also, both
sneakers and loafers become treacherous on even slightly muddy surfaces. The Vietnam combat boots (see Figure 7) excel in many ways.
In the first place, they are not too expensive (about twelve dollars).

They have eyelet holes around the soles which drain any water that may get into the boots, and they have deep-cleated rubber soles which give superior traction on muddy surfaces. These boots, however, are becoming increasingly difficult to find, so you may have to buy another type. Work boots with heavy, deep-cleated rubber soles will suffice. Buy boots with eyelet holes for the shoestrings rather than the hook-type lacings which are found on most hiking and climbing boots. Hooks catch on the cable in cable ladders and are almost impossible to remove.

Your hard hat should be the best you can afford. A rock climber's helmet (see Figure 8), with a foam lining and fitted with a lamp bracket, is the safest and most comfortable hat you can buy. It is not easily knocked off your head in a fall and has very high impact resistance. These features give a rock climber's helmet a large safety advantage

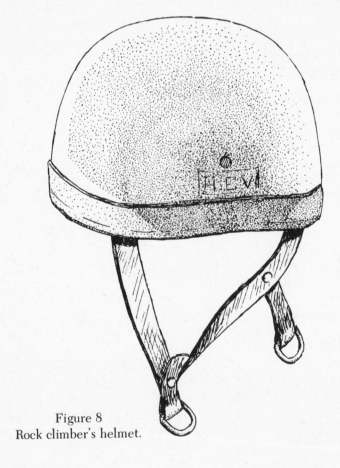

Figure 8
Rock climber's helmet.

over the suspension helmet. The principal disadvantage of this superior helmet is its relatively high cost (about fifteen dollars). For this reason most cavers stick to suspension helmets. Provided that you avoid standing around the bottom of drops, where something or someone may fall on you, you will have a good measure of safety with a suspension helmet—as long as it meets the minimum safety requirements. Such a helmet (see Figure 9) will be marked "Approved for high-impact safety. . . ." Another safety requirement is the suspension system itself. This should be adjusted so that there is a distance of 1½ inch between

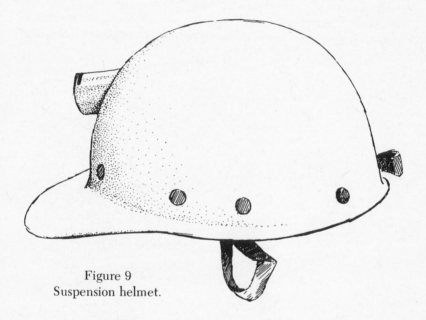

Figure 9
Suspension helmet.

the crown of your head and the top of the helmet. This should be done with the horizontal adjustment so that the helmet doesn't slip forward onto your nose when it is carrying a carbide lamp on its bracket. Neither should the helmet be so tight that it is uncomfortable. Look at the suspension system before you buy. If it has numerous holes so that fine vertical and horizontal adjustment is possible and if it bears the "High-Impact Safety" sticker, it should prove satisfactory for caving. Remember to always wear a chin strap. The elastic type is most comfortable.

The Cave Pack

What you carry in your cave pack will depend on the kind of cave you are about to explore and the length of time you plan to be underground, although the "Essential Items" should, of course, be in your pack for every trip.

The pack itself (see Figure 10) can be an inexpensive Army surplus canvas shoulder pack. But only the type with strap and buckle closures. Snap closures become useless when impregnated with mud. Many cavers prefer to use an Army ammunition box as a pack if they are carrying cameras or other delicate items because the ammunition box is both waterproof and protectively rigid.

The contents of the cave pack can be divided into essential items and optional items.

Figure 10

Small canvas shoulder pack with strap and buckle closures is ideal for caving.

ESSENTIAL

Spare carbide in one or more plastic baby bottles or other waterproof, unbreakable container. (12 hours of caving will require about i lb. of carbide.)

Water. Drinking water and carbide-lamp water can be carried in the same container. Avoid drinking cave water and don't use it in your lamp unless this is unavoidable. Cave water is frequently polluted and can make you sick. Also it usually contains a fine mud suspension which gives it an evil taste and which will clog your carbide-lamp water dropper. Two pints of water should be sufficient for a 12-hour trip both for drinking and for your lamp. Carry the water in one or more plastic containers in your pack.

Flashlight. As long as it is a reliable flashlight, any type will do. The most useful type is a rubber-encased waterproof flashlight with a loop at one end. In many instances it is reassuring to have the loop attached to the carabiner at your belt. In such a position, it is accessible if your lamp goes out. This will frequently happen in pits when the ever-present spray or drips of water extinguish the carbide lamp while the caver is climbing either up or down.

Spare batteries and *spare bulb* for flashlight.

About 6 inches of candle.

Matches in a waterproof container.

Spare parts for carbide lamp (spare gasket, felt, and 2 tips).

Tip reamer (the brush type is best). It can be carried on a 6-inch loop of string and attached to the suspension of your hat.

Pencil and paper, useful in emergencies.

NOTE: These last six items should be carried in a waterproof, unbreakable container.

Small pliers.

Food. This should be well protected from cave dirt and water in an unbreakable container.

Plastic bag and elastic band closure for spent carbide.

OPTIONAL

First-aid kit in waterproof container.

Aluminum cup.

Solid Sterno fuel.

Spare sweater.

Whistle (on cord around your neck).

Vertical equipment (in separate pack).

First-aid kit

An advanced first-aid course is a must for cavers. The contents of your first-aid kit should be relevant to your first-aid qualifications. If you are qualified you will be able to carry and use a variety of drugs and supplies. If you are unqualified and know little or nothing of first aid, the following will be adequate for your kit: Band-Aids, aspirin, sterile gauze pads, antiseptic cream, and a large square of clean cotton cloth (about 24 inches square). While not normally part of a first-aid kit, such a piece of cloth can serve as a makeshift sling, a bandage, an absorbent for bleeding, or a pad, among its many other uses.

Food

Individual eating patterns will largely determine the choice of food. The type of cave and the length of time you will be underground are also important factors. *You should always carry food whenever you are underground.* It is a source of quick energy, and can be very important in some emergencies. For a short (less than 4 hours) dry trip, no more than a snack is needed. Your choice should be made from high-energy uncrushable food. An assortment of pitted dates, raisins, candy, Space Bars, and unsalted nuts makes an ideal snack. Avoid anything that is salty. The development of a severe thirst while underground will take much of the pleasure out of a trip and may achieve the proportions of a disaster. In desperate thirst, people have been known to drink the water in their lamp and all the water they were carrying for the lamp. For long or wet trips, take food that can be easily heated. Quick, warm energizers can be made by dissolving a bouillon cube in water in a tin cup, then heating it over a carbide lamp flame or a can of Sterno. Chocolate dissolved in hot water or cocoa mix in hot water is also very pleasant. If your trip is sufficiently long to encompass a real mealtime, small cans of beef stew can be satisfactorily heated (after opening) on your can of Sterno. Small cans of pudding and even black bread and cheese complete the meal. There is no real weight advantage in carrying dehydrated food into a cave in preference to canned food. The saving in weight becomes negligible when you consider all the extra water you have to carry to reconstitute the food.

Light Sources

Three separate sources of lighting should always be carried. The carbide lamp (see Figure 11) is, of course, your chief reliance, with the flashlight and the candle being the backup sources. If you are

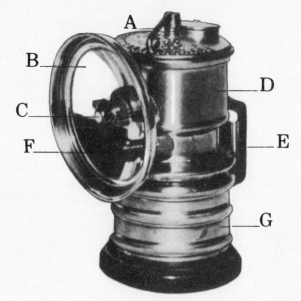

Figure 11
Premier hat lamp

A. Lever for water-drip control
B. Reflector
C. Ceramic tip
D. Water chamber
E. Hook which attaches to bracket on hat
F. Flint lighter
G. Carbide chamber

left without light in a cave and it so happens that the others in the group have expended their light, your chances of survival would not interest a betting man. With this in mind you should aim for more than enough carbide light for the particular cave you have in mind and for thoroughly reliable spare sources of light. It should be unnecessary to point out that your carbide light may be useless if it is not kept in excellent working condition *always*.

Carbide light is the prime choice of cavers throughout the world. The light is of the diffuse kind that is nonetheless very bright, and it is the cheapest type of light to produce. Both of these factors make it particularly suitable for caving. It does have some limitations, however. It becomes a nuisance in wet caves. At such times, an electric system for the primary light source has obvious advantages. However, along with wet suits, electric systems are initially very expensive and are a luxury beyond the needs of basic caving equipment unless your pocketbook is fat and/or you intend to specialize in wet caving. Carbide-light users generally get around the wet spots in caves by turning out their lamps and using their waterproof flashlights. The lamps can be relit afterwards with the flint lighters which are mounted on the reflectors. A periodic check should be made on the condition of the flint in the lighter. If it is beginning to wear down, replace it. There is a useful trick to relighting the lamp with the flint: Cup your hand over the reflector so that some gas accumulates between hand and reflector, then quickly run the same hand down over the flint wheel. With luck, the spark will ignite the accumulated gas.

A small brass lamp that can be mounted on the hat is the most popular type of carbide lamp in caving circles. There are two brands available: the Justrite brass lamp, and the Premier brass lamp. They are comparable in quality and performance. However, Justrite appears to be phasing out production of the metal lamp and replacing it with a plastic type. The latter is not very practical for caving. Cavers are having an increasingly difficult time buying the metal Justrite, but fortunately the Premier lamps are becoming more readily available in the United States.

Kneepads

Kneepads are essential in most caves. Crawling on hands and knees through small passages without pads causes unnecessary pain and sometimes permanent injury to the knees. Either athletic kneepads or workman's kneepads are suitable. Both types are generally available at

Army surplus stores. Athletic kneepads are constructed of foam and elastic and are worn next to the skin (over liberal quantities of talcum if the trip will be a long one). They are not bulky and allow freedom of movement during climbing and walking as well as when crawling. Correctly fitted, they are tight enough to stay in place when you are moving, but loose enough to allow free movement. Note that if they are too tight they can cause swelling of the legs and other circulation problems. Workman's kneepads are black molded-rubber shields that buckle on over pants. They are bulky and clumsy. However, they can be quickly removed and put in a pack when not in use in a crawl way.

3.   Choosing and Maintaining Caving Equipment

For underground safety, cavers rely on the quality of their equipment. To be precise, they rely on quality equipment in first-class condition. A reputable dealer in caving and climbing supplies will provide you with good quality ropes and caving hardware. It is his business to sell it to you in good condition, but it is entirely up to you to properly maintain its condition. Good equipment maintenance is more than protecting your investment, it is protecting your life.

Selecting Equipment

CARABINERS

The spare carabiner on the belt-handline rig and the carabiners used in conjunction with brake bars in your tandem brake-bar rig are nonlocking. For all other purposes in caving, locking carabiners should be used. The lock adds a margin of safety at any point where the shock of a fall will be taken predominantly by the carabiner. The purpose of the screw lock is to prevent the gate from opening unless manually opened (see Figure 12).

Both aluminum and steel carabiners are acceptable for caving. The former are lighter in weight, but are more expensive than steel carabiners. In terms of strength there is little to choose between them, so your choice can be directly based on the amount of money you are able to spend.

Choosing between the various brands of carabiners that are on the market is bewildering because so many appear to be of comparable strength, quality, and cost. My only suggestion is that you buy carabiners of the more popular makes. Stubai, Bonati, and Recreational Equipment, Inc. (R.E.I.), are three well-known brands. Generally speaking, popularity doesn't go unearned.

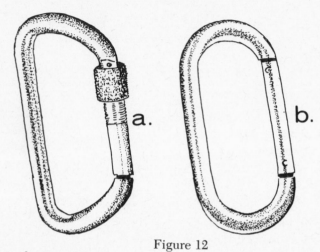

Figure 12
a. Locking carabiner   b. Nonlocking oval carabiner

In the last few years more care has been exercised to keep carabiners of dubious quality off the market. Therefore, your chances of buying an intrinsically unsafe or a faulty carabiner are slim, but you should still examine any carabiner that you buy before you use it. Look at it with a hand lens. Watch for corrosion marks and pitting. Examine the gate and hinge pin to look for faulty alignment. Screw up the lock on the gate. It should not become unscrewed by itself, so shake it around and check for this. If you have any doubt about anything you find, send it back to your supplier and ask for a replacement.

BRAKE BARS AND LINK FOR
TANDEM BRAKE BAR RIG

Although adequate strength and good quality should be looked for when buying brake bars, the most important thing to consider is how well the brake bars fit the nonlocking carabiners on which they have been placed. On a well-matched carabiner and brake-bar set, the brake bar should be easily movable up and down the carabiner when it is closed, but it should not move by itself. Both R.E.I. and C.M.I. (Colorado Mountaineering, Inc.) make brake bars that

Figure 13
Brake bar.

fit snugly on their own carabiners. A regular, heavy chain link can be used to attach one carabiner to the other. A typical brake bar is illustrated in Figure 13.

PULLEYS

A small rescue pulley (see Figure 14) should be part of every caver's equipment once he has graduated beyond completely horizontal caves. R.E.I. makes an inexpensive rescue pulley that has adequate strength. The Dolt pulley is more expensive but is possibly the best quality pulley available. A rescue pulley is part of the equipment used in the practice system detailed in Chapter 9.

Figure 14
Rescue pulley.

ROPES

Available ropes can be divided into three categories based on construction (see Figure 15), which in turn are available in five varieties as sold on the market:

1. Laid, natural fiber rope.
2. Laid, synthetic fiber rope.
3. Dynamic braid rope.
4. Low-stretch braid rope.
5. Webbing rope.

1. *Laid, natural fiber rope.* The use of rope from this category—Manila rope—has become obsolete in caving. Natural fibers lack strength and tend to rot in the damp cave environment.

2. *Laid, synthetic fiber rope.* In this category, Plymouth Goldline (the tight mountain-climbing lay) is the best-known representative. It is versatile; it can be used for belaying and for rappelling and Prusiking, and it is inexpensive. It has two annoying drawbacks, however. Because it is a laid rope, it produces an uncomfortable spin when used for such climbing techniques as rappelling and Prusiking. Because of the spin, many people experience both vertigo and nausea when they are climbing up or down rope. The second disadvantage is the stretchiness of the rope. The stretch makes it an acceptable belay rope but it is inconvenient otherwise. When Prusiking, the

climber's weight stretches the rope by approximately 15 feet on a 100-foot drop so that he will have to climb 15 additional feet because of the stretch. Goldline could be classed as a satisfactory rope for all-purpose caving. It is available in almost any length wanted or needed.

3. *Dynamic braid rope.* Of this third category, Eldelrid Perlon is the best-known representative. Mammute Dynamic and Edelweiss are two other brands that have recently become popular. Eldelrid Perlon, however, comes closer to being the ideal all-around caving rope. It is pliable when new and remains pliable after years of use. Most other types of rope become stiff and unwieldy as they age. It is the epitome of a good belay rope and also provides a very smooth, spin-free rappel and Prusik. It has up to 50 per cent stretch when shock loaded and this great stretch factor gives a dynamic quality to even a static belay. When not shock loaded, or if it is loaded with less than 200 pounds, the effect of the intrinsic elasticity is barely felt. On a 100-foot Prusik, the climber's weight on the rope would cause it to stretch no more than about 12 inches. The braid construction is the reason that it does not produce spinning. The one disadvantage is the cost. It is nearly twice as expensive as either Goldline or Bluewater, and it comes only in a standard 165-foot length. Samson 2 in 1 braid rope is

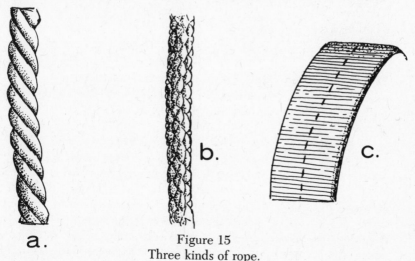

a.

b.

c.

Figure 15
Three kinds of rope.
a. Laid, synthetic-fiber rope.   b. Braid, synthetic-fiber rope.
c. Tubular nylon webbing.

not a good choice for caving. The sheath is too loosely braided over the core and this has caused some unpleasant caving accidents.

4. *Low-stretch braid rope.* Bluewater II is a good example of rope from the fourth category. It has become famous for being the first rope constructed exclusively for rappelling and Prusiking. It is a tight braid rope, of excellent quality, that gives a spin-free climb. It is inexpensive and can be bought in almost any length. It has almost no stretch, however, which makes it useless as a belay rope and therefore useless as an all-purpose caving rope.

5. *Webbing rope.* Tubular nylon webbing is, for caving, the most useful member of this fifth category of ropes. It is strong and can be used for a seat sling, chest harness, handline, or sling for tying off cable ladders and pulleys. Recreational Equipment, Inc., retails a very popular 1-inch tubular nylon. Automotive safety-belt webbing, 2 inches wide, makes very comfortable seat slings for rappelling.

All the ropes mentioned in the above categories have been used in caving for years. With the exceptions of Manila rope and Samson 2 in 1 rope, they are proven, good quality ropes with a useful strength of 4000 pounds or greater. This 4000-lb. strength is quite adequate for caving purposes.

For your first caving rope, I would recommend a 165-foot length of Perlon, if you can afford it, or a 150-foot length of Goldline. A length of Bluewater II rope can be added to your equipment if, after you have been caving for some time, you find need for an extra, strictly rappelling and Prusiking rope. For belaying, rappelling, and Prusiking, don't buy rope of less than 7/16-inch or 11-mm. diameter.

Carefully examine every inch of your rope after you buy it and look for signs of damage to the outside fibers or to the core. Core damage can be recognized by spots that are soft and spongy when squeezed, or by lumpiness. Return any doubtful rope to the supplier. For all practical purposes, the useful strength of the rope can be considered to be halved if the rope has ever been used to arrest a caver's long fall, or if it is abraded.

Remember that the strength of two lengths of rope joined together by a knot or knots is considerably less than the strength of an unknotted length. Avoid climbing on a joined length and *never* belay with a knotted length.

Caring for Equipment

ROPES AND SLINGS

Synthetic fibers have replaced natural fibers in cave ropes. Generally speaking, they are resistant to the detrimental effects of cave mud and water, but they still need special care. The following dos and don'ts apply to tubular nylon webbing as well as to braid and laid ropes.

Don't leave rope lying in the sunlight. The ultraviolet rays of the sun have a detrimental effect on most synthetic fibers.

Don't stand on the rope.

Do keep the rope clean. When it becomes muddy or gritty, wash it in a washing machine with mild soap. Ivory Flakes are suitable. Don't use detergents which contain bleach. The bleach weakens synthetic fibers. Longer ropes can be washed in the jumbo-sized washer at the laundromat.

Do pad the ledge underneath the rope at the top of the pit. If this is not done, the ledge will abrade the rope. If a long fall is taken on the rope, the rope may be cut in two where it passes over the unpadded edge. An abraded rope is a weakened rope and should be treated the way fall-weakened ropes are treated. However, if ledges are padded, the chance of abrasion is slight. Frequently check every inch of your rope for damage.

Do carry the rope through the cave in a bag whenever possible. This avoids abrasion, snagging, and getting mud on the rope. A clean rope is easier to use than a mud-covered one. A duffel bag is excellent for this purpose, and easily doubles as a rope pad at ledges. It can be kept in place at the ledge by anchoring it to a tree or rock with a length of parachute chord.

Don't store ropes where they can be splashed with cleaning fluid or chemicals. They should be hung in a clean, dry place.

All cut ends of synthetic rope and sling material should be melted with a cigarette lighter to fuse stray fibers. After this they should be whipped with carpet thread (see Figure 16). It is important that this be done, in order to prevent any subsequent fraying.

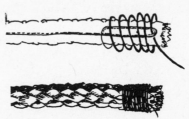

Figure 16
Treated rope end.

Do assume, if a long fall is taken on a rope, that the rope is permanently damaged. It should be retired as a belay rope, cut into shorter. lengths for handlines or used for hauling equipment in and out of pits.

Figure 17 illustrates one of several satisfactory methods of coiling a rope. A coiled rope is much easier to handle when it comes time to use it at a drop than is a tangled one.

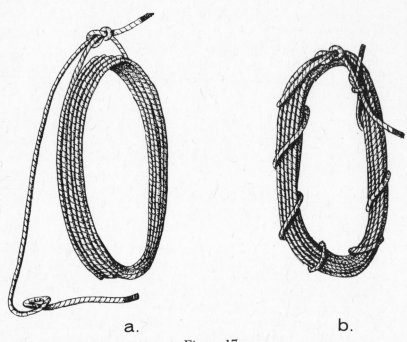

a.   b.

Figure 17
Recommended method of coiling rope.
a. Coil rope leaving a long end and a short end, then tie a bowline in the two ends.
b. Wrap the long end around the coil and tie a second bowline.

CARABINERS AND OTHER HARDWARE

Carabiners should be washed and dried after any trip when they have gotten either dirty or wet. Aluminum carabiners are less susceptible to rust and corrosion than are steel carabiners, but even these should be kept clean, because clean locks and gates are easier to manipulate than are muddy or rusty ones. If hardware is to be stored for a long period of time (six months or longer), it should first be lightly oiled. The oil can be wiped off when next you are ready to use the equipment.

CABLE LADDERS

Cable ladders should never be left in a cave between trips. They are susceptible to damage caused by humidity. Otherwise they should be cared for by the method outlined above for carabiners.

The Carbide Lamp

Here are a few suggestions for getting the best service out of your carbide lamp. An unreliable lamp is frustrating to use and dangerous because, in caving, your safety depends heavily on good light. Therefore, the few minutes spent on the care and upkeep of your carbide light will have been well spent when you get into the depths of a cave.

1. Avoid using cave water to fill your lamp. There are usually mud particles suspended in cave water and this clogs the water dropper of the lamp.

2. When the flame dies down after you have finished using your lamp, turn the water off and blow out the flame. Allowing the flame to die out after use causes the tip to become clogged with carbon.

3. Use good quality ½-inch lumps of carbide.

4. Clean out spent carbide as soon as you finish using the lamp. If you leave this carbide in until the next trip, corrosion around the thread seal will make it very difficult to unscrew the top from the bottom of the lamp when you need to refill it.

5. A shiny reflector gives better light. Toothpaste contains a fine abrasive which is an excellent reflector cleaner and polisher.

6. Wash out the lamp base and felt, polish the reflector, and flush out the top of the lamp with water until the water is freely dropping from the stem, after every trip.

OPERATION OF THE LAMP

Set valve lever at OFF position.

Fill top half of lamp with clean water.

Half fill the base with carbide.

Turn lever ON about three notches.

Wait a few seconds after turning lamp on and then light the gas. It takes time for the flame size to catch up with the increase or decrease in water flow, so move lever only one notch at a time and watch flame until it has responded to the change in flow before moving the lever further.

CARBIDE LAMP PROBLEMS

Most of the irritations associated with carbide lamps can be avoided if you observe the correct procedures for the care and operation of your lamp. The principal difficulties that most cavers experience with their lamps are along the following lines:

Intermittent, spluttering flame. This indicates that the base is flooded. Turn off the lamp and blow out the flame. Put your mouth to the tip and blow out any water in the gas line. Unscrew the base and tip out any water that may be there. Screw up lamp, then relight it.

Flame around the seal. This usually indicates that the base is not tightly screwed onto the top. Blow out the flame around the seal and blow out the tip flame. Screw the base onto the top more securely and relight the lamp. If the seal ignites after this, it is likely that the rubber gasket is old and cracked. Replace it with a new one. Keep the thread clean. A clean thread provides a better seal.

Clogged gas line. Any part of the gas line can become clogged. A sodden felt will prevent the passage of gas. Carbon clogging at the tip is also an offender. The first problem can be rectified by replacing the wet felt with a dry one, and the clogged tip can be reamed.

Most brands of carbide lamp have a tip that slides into the mounting. With this type, the tip is gently tapped with a heavy implement until it is firmly in place. Less common is the lamp with a screw-threaded tip and tip mounting. With both types of tip small pliers are a useful accessory when it becomes necessary to remove or replace the tip.

4.   Essential Knots for Caving

Knots, and the ability to tie them correctly and to use them proper-ly, are an important part of caving and of caving safety. With one or two exceptions, there is nothing particularly unique about these knots. For the most part, they are the kinds of knots one would encounter almost anywhere. A bowline is a bowline, when all is said and done, and so is a half hitch. Still, there are half a dozen or so knots which are absolutely essential to the caving enthusiast. No book on caving would be complete without a brief survey of these basic knots. If fur-ther information is needed, it can be found in any one of many books and manuals on knots and their use.

Single Bowline
This knot is used for tying off the rope for belaying, for a handline, or for a fixed rope for rappelling and Prusiking. It can be tied at the end of the rope with a single thickness or into any part of the middle of the rope with a double thickness, as shown in the accompanying illustration.

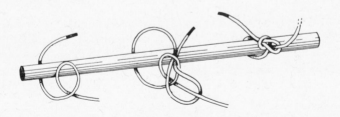

Figure 18
Single bowline.

41

Bowline on a Coil

This knot is used in addition to two wraps of rope around the chest, for the climber's belay loop. The bowline should be tied into both coils and should be finished with a half hitch or overhand knot.

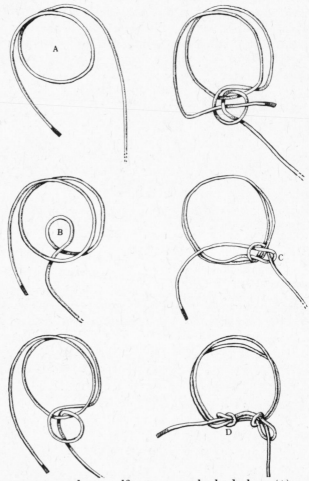

1. Wrap rope around yourself twice to make body loop (A).
2. Make bowline loop (B) by making a small clockwise loop in the long end of the rope.
3. Flip bowline loop outward until it is resting on long end of rope.
4. Pass short end in and out of bowline loop, under the long end.
5. Loosely tied bowline (C).
6. Tightened, finished bowline with overhand knot (D) as back-up knot.

Half Hitch, Figure 8 and Overhand Knots

The half hitch is never used alone. Its use is confined to finishing off other types of knots.

The Figure 8 and the overhand are both loop knots. They are used primarily for tying equipment onto a rope for hauling purposes. The Figure 8 loop is sometimes used for the belayer's loop rather than the Butterfly knot. This practice is not recommended because the Figure 8 knot has a tendency to bind up and become impossible to untie if a sudden stress is placed on it. The overhand knot is often used to finish off a bowline.

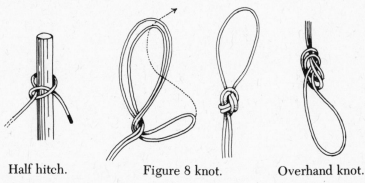

Half hitch.   Figure 8 knot.   Overhand knot.

Butterfly Loop

This is the loop that is recommended for the belayer's loop. It is the strongest of the loop knots and is the easiest to untie after it has been stressed.

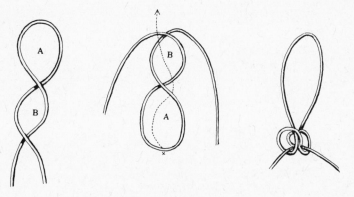

Butterfly knot.

Fisherman's Knot

This knot is used for joining two ropes of equal or near equal size. Its greatest use is in the construction of slings.

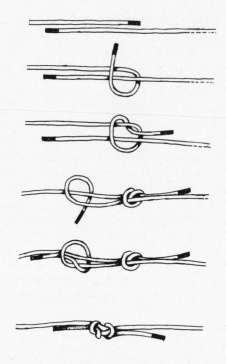

Figure 24
Fisherman's knot.

Prusik Knot

The uses of this knot are described in the chapter on Prusiking.

5.   General Caving Safety

Leadership

Be realistic about the capabilities of the individuals in the caving party and the amount and type of equipment that is available to the group when choosing caves to visit in your early caving trips. Explore well-known caves for a while because these have usually been mapped. A map provides valuable information about the nature of the cave. Length and ceiling height of passages, water traps, depth of pits, squeezes, and floor breakdown—these are the types of features that are noted on the map. More particular details can be obtained by asking cavers who have visited the cave. It is necessary to get an accurate indication of the need for any special equipment such as ropes, handlines, hardware etc. Do not attempt anything for which you are inadequately equipped. Begin your caving with easy caves and work up to the difficult ones.

The above paragraph was written mostly for the benefit of cavers who do not have access to an organized caving group and therefore begin their caving career without the guidance of experienced cavers. Write to the National Speleological Society (N.S.S.), Cave Avenue, Huntsville, Alabama 35810, for a list of the Society's Grottos (chapters) and if there is a Grotto in your home area or the area in which you wish to cave, join it. Most caving accidents involve cavers who are attempting something beyond their level of experience and expertise. Special techniques should be learnt outside the cave. Joining a Grotto is no guarantee of safety but it generally provides more safety than caving with a maverick group. In a Grotto, you will be caving with experienced cavers from your first trip on. Most Grottos offer some practical training, have cave maps and publications available for borrowing, and will have a pool of equipment and expertise greater than can be found in a group of casual cavers. All of these are excellent reasons for joining a Grotto.

Duties of the Leader

The leader is the person who makes the ultimate decisions. The feelings and opinions of every member of the group should be consulted when decisions have to be reached, but it definitely reduces potential friction, avoids confusion, and saves time to have only one "captain." This does not mean that individuals can knowingly neglect personal safety precautions and expect the leader to remedy every mishap. Ideally, the leader is the most experienced caver in the group and should be willing to accept as his own the following responsibilities:

1. Inquire into the level of experience and adequacy of personal equipment of each individual before the trip and be firm about not allowing the participation of people whose experience and equipment he considers inadequate for the selected cave.

2. Find out as much as possible about the cave to be explored and make sure that any special equipment that is needed goes underground with the group.

3. Make sure that at least one list of the participants is left with a responsible person.

4. Check the weather forecast.

5. Closely watch the performance of novices. If they need help, make sure that someone supplies it. If they are having trouble keeping up with the group, slow down. Watch for signs of exhaustion, hypothermia, panic, or other problems that may be difficult to reverse if not noted early. It may be necessary to check the actions of cavers who are behaving in a careless or thoughtless manner if their actions could jeopardize their own safety or the safety of other members of the group.

6. Although everyone should be taking note of the route, route-finding is the ultimate responsibility of the trip leader.

7. The leader or an experienced person chosen by him should check all riggings of ropes, belays, Prusik harness and slings, and rappel equipment.

Leading a trip is a great responsibility and should always be taken seriously. A good leader is adept at avoiding mishaps, but also knows where and how to begin to rectify them should they occur. Cavers have accepted the leader's authority if they participate on his trip.

Size of the Group

The minimum number of people on a caving trip should be four. In the event of an accident to one of the group, there will be one person available to remain with the victim, and two to go for help. A group of four means that neither the victim nor the person going for help need be alone. Lone caving is unsafe and seldom justifiable. A group of four to six people is the ideal size for most purposes. A group of this size is able to move through the cave quickly and efficiently without bottlenecks at squeezes and pits. A large group is not necessarily unsafe but is always slow. On expedition-type trips into particularly difficult caves, however, a large group is often needed to transport the mass of necessary equipment and supplies.

Direction Finding in a Cave

Direction finding should not be only the responsibility of the leader. With all members of the party sharing this task, getting lost or getting temporarily mislaid in a cave should be next to impossible. Even after a particular cave has become familiar to some of the party so that they are able to relax their usual route-finding methods, any newcomers along on the trip should be encouraged to route-find in the usual manner rather than follow one who happens to know where he is going.

Direction finding in a cave depends heavily on a retentive memory, although there are several memory aids which can be used. Route-finding is a survival skill and should be practiced from your very first trip underground. It should be developed into a habit. Although a map of the cave is a very useful accessory, especially if you are undertaking a systematic exploration of a cave, it should be considered only an accessory, not a replacement for careful observation as a means of finding your way around.

Basic route-finding consists of little more than noting the more prominent features, *e.g.*, an oddly shaped or particularly large stalactite, passage or cavern cross sections, and wall and ceiling features. However, the most important aspect of this method is frequently to turn around and examine the recently traveled passage as it appears looking back. This is the way it will appear on your return journey. The difference in appearance in the same passage or cavern from opposite views is remarkable. This process of memorization becomes particularly important when you get to a junction of several passages. At this point you should come out of the passage into the junction area, then examine the whole character of the junction and, in particular, the relationship of the entrance of the passage you just exited to the rest of the junction area. Then note the relationship of the one you are about to enter to the one you just exited.

If you feel you need extra help with route-finding, use small squares of aluminum foil, each with a 1-inch square of reflector tape stuck on its center, as route markers. These should be removed on the way out of the cave.

Miscellaneous Cautions

As you travel through the cave, always evaluate the obstacles with the return journey in mind. Negotiating an obstacle from the opposite direction may require different skills, perhaps more skill. It will sometimes require different equipment. Be prepared for this.

Many well-known non-tourist caves contain old wooden ladders, poles, and ropes, placed there by well-meaning cavers. There is no way of knowing how long they have been in the cave and timber and ropes deteriorate quickly in the cave environment, so don't trust these things. Metal climbing aids that have been left in the cave fare better than timber, but still deteriorate over a long period of time, so it is better not to trust these, either.

Appropriate equipment and skill are only part of the safety requirements in cave exploration. The other part is you. Good physical condition is advisable in horizontal caves and an absolute necessity in vertical caves. The misery of ascending a 200-foot pit with only enough energy for the first 100 feet can well be imagined. Exhaustion and fatigue make people mistake and accident prone.

6.   Moving Through a Cave

A passage large enough to walk along occasionally appears in a cave. It usually alternates with other passages that dictate any one of several modes of travel. It is possible to generalize about the best way to move through the cave that covers many of these modes: Watch where you put your feet. Cave rocks are usually slick with mud, and rock breakdown on the floor of the passage will probably be unstable as well as slick. For the same reasons, don't leap from rock to rock. Test dubious footholds and handholds before putting your weight on them. Carry your carbide light in your hand whenever both hands are not needed for climbing. You will get a better look at what you are traveling over with a hand-held lamp. Move rhythmically to conserve energy. Set a steady pace and avoid stopping and starting any more than is necessary.

The means of getting around a cave—stopping, crawling, walking, rock hopping and crouching—need no explanation. But squeezing, rock climbing, and chimneying represent more difficult caving and sometimes cause problems, so these will be treated separately.

Squeezing

Squeezes come in an impressive variety of types. By definition, a squeeze is a tight, confining passage but it can be corkscrew shaped, "S" shaped, "Z" shaped, bell shaped, vertical, horizontal, a crevice, or any combination of these types.

Horizontal straight squeezes seldom cause problems. In a straight tube, if you can get your head through with your hardhat on, you should need little more than determination and patience to get the rest of your body through. If there are sharp and extreme bends in the squeeze, however, it will not be long before impossible demands will be made on the bending capacity of bones. Such bent squeezes are sometimes impassable, not because the passage is too tight for human entry, but because the bends are too extreme. Travel through the passable ones with as little as possible of the length of your body exposed to the bends. Try moving through upright, on your knees, if possible.

Canyon or keyhole squeezes are crevices that are traversed with the body held lengthwise, upright, or anywhere in between. It is an advantage in this type of squeeze to select the widest place about three or four feet ahead of you, move toward it, and then repeat the process until the squeeze has been exited. You will need to develop the ability to stay in the squeeze at the level that you have selected. It can be inconvenient or even disastrous to slip further down in the crack where it may well be narrower.

A straight vertical squeeze will seldom present any problem when you are descending. You just lower yourself, inch by inch, with gravity to help. However, ascending the same squeeze may be a nightmare because the narrowness will prevent you from raising your limbs into a position where you can exert upward leverage. Bell-shaped vertical squeezes are also uncomplicated to descend but can be difficult or even impossible to ascend because the squeeze bells outward, away from the reach of the feet, frequently too far away to present rocky projections from which you can push up. Figure 25 illustrates a bell-shaped squeeze with a straight vertical squeeze above it. If you mentally put yourself in the position of the ascending caver, you can imagine how impossible it would be to ascend without the help of a handline.

Safety in Squeezes

A squeeze which presents the combination of a crevice and tight bends is probably the most difficult and aggravating type. Next would be a vertical squeeze with vertical bends. Treat these with respect and avoid them until you have done several easier ones to test your reaction to confined, difficult situations, and until you have built up some skill and self-confidence. Knowing which squeezes are possible for you to do, and which are either impossible or too dangerous, lies in having prior knowledge of the particular cave in question, or in having enough caving experience to accurately anticipate the difficulties of a particular squeeze and your own ability to handle it. Be particularly careful to anticipate the difficulties of the return journey through the squeeze, in addition to those encountered going in. As a novice your best protection is to stick to well-known caves for a while—preferably caves with a map that you can look at. Bad squeezes frequently have some fame attached to them.

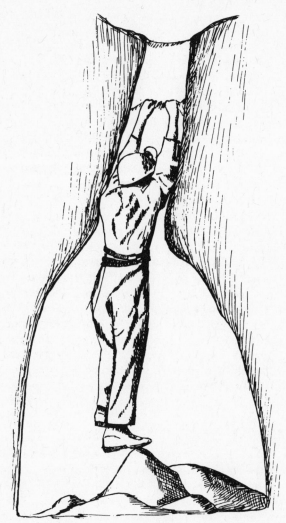

Figure 25
Combination vertical squeeze and bell-shaped squeeze.

Always rig a handline for use in any straight or bell-shaped vertical squeeze. This may not be needed for the descent, but it will be absolutely necessary in order to haul yourself up when you ascend.

Do not enter a downward sloping squeeze headfirst. It is undeniably easier to negotiate such a squeeze headfirst and you can see where you are going, but if you become temporarily stuck you face the possibility of blacking out if your head is too much lower than your feet.

In addition, backing out of a steep downsloping squeeze, if that becomes necessary, will be difficult or even impossible.

Whenever possible, it is better to move through a squeeze with both arms in front of you for three reasons: (a) the chance that one arm will be pinioned uselessly under you is eliminated, (b) it reduces the diameter of your body in a very tight place to align the arms alongside the head rather than alongside the comparatively thick chest and torso (c) when toeholds are absent, especially in a vertical squeeze, the strength of both arms and use of all ten fingers may be necessary to pull yourself up. The strength of one arm in a restricted place is not enough to lift your body weight.

Always push your cave pack ahead of you. It is easy to imagine squeeze situations in which keeping the pack alongside or behind you could be a nuisance.

If you have taken the precaution of first putting a handline down a vertical squeeze and keep the other cautions in mind, it is unlikely that you will get stuck in a squeeze. It sometimes happens, however, to even the best and most experienced cavers. In such a case, first face the fact that you will have to get yourself out. It is seldom that companions can get close enough to effectively help you. Your greatest assets in this situation will be a cool head and determination. You cannot give up. Back off a little, redistribute some flesh, give a timely wriggle, then try again. Talk to your companions in order to help yourself stay calm. Panic uses up valuable energy, so fight it.

Like everything else in caving, the process of moving through a cave is not so simple as it may sound. It has its own techniques and procedures. If they are learned carefully at the outset, the adventure of caving will be made happier, and safer and more rewarding.

Rock-Climbing in a Cave

This is best learned with a belay outside the cave. Simple rock-climbing involves keeping the weight over the feet (don't lean into the rocks), using the legs to climb and the arms for balance, and taking small steps. Anything beyond this basic knowledge should be learned from a skilled rock climber, or a good mountaineering handbook. It is the size and availability of handholds and footholds which determines the ease or difficulty of any climb.

Chimneying

This is a counterforce technique of rock-climbing used to climb up or down between two parallel rock walls. It can be practiced in and demonstrated in a doorway. (See Figure 26.)

Figure 26
Chimneying Technique.

7. Belaying

Belaying, simply stated, is a technique whereby one climber, securely positioned at either the top or the bottom of a drop, is able, through a system of ropes, to arrest the fall of a fellow climber.

When to Belay

Belay any climb where handholds or footholds are scarce or where difficult moves have to be accomplished regardless of the length of the climb. Belay every climb where the climber could fall ten feet or more. Do not substitute a handline for a belay unless the climb is very short and very easy.

Caves generally dictate a moderately fixed belaying routine. As one progresses into almost any cave the climbs initially encountered require down climbing. In such cases the party usually will pause at the drop, a belay position for a top rope belay is set up (see Figure 27), and then each climber except the last is belayed down the climb. On expedition or assault trips into the more formidable caves, or on drops where double rope belaying requires an unavailable length of rope, it is common to leave this last person at the top as a belayer. He remains there until the group comes back up the drop on the way out. Such belayers form part of the support team. In lesser caves, after the other members of the party have been safely belayed down the climb, the most experienced caver would rig a double rope belay (see Figure 28). He lowers one end of the rope to the group below, then ties the other end around himself in a bowline on the coil. The slack is taken up, and he is then belayed by someone at the bottom as he climbs down. In the same way, on the way out of the cave, the most experienced caver climbs up first, belayed from the bottom with the double rope belay which has been left in place after the down climb, resets the rope for a top rope belay, and then top belays the rest of the group.

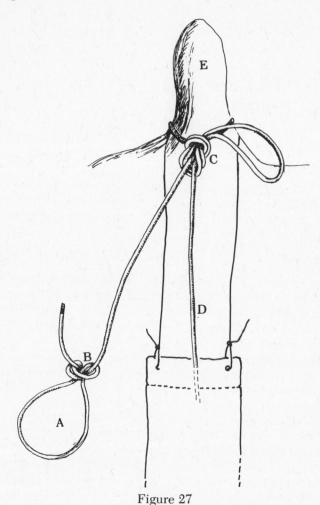

Figure 27
Top rope belay rigging.

A. Belayer's loop.   D. Rope to climber via belayer.
B. Butterfly knot.   E. Sturdy anchor point.
        C. Single bowline tied with doubled end of rope.

How to Position the Belay

Assuming that the climber is going to fall will help you to decide how and when to set up the belay. This assumption will make it less likely that the belayer will take a nap while belaying. Too many cavers

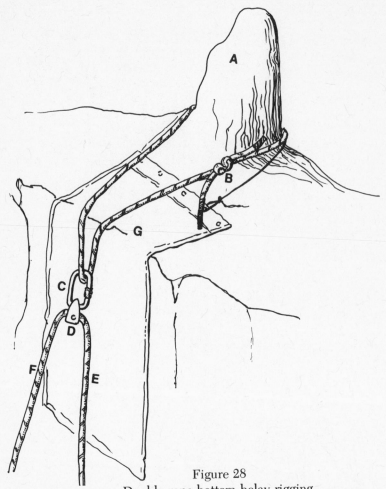

Figure 28
Double rope bottom belay rigging.

A. Stout anchor point.
B. Fisherman's knot.
C. Locking carabiner.
D. Rescue pulley.

E. Rope to the climber.
F. Rope to the belayer.
G. Rope pad.

think of their turn to belay as a rest stop. The belayer must remain constantly alert to interpret any changes in tension in the rope, and to respond to any requests made by the climber.

To position the belay correctly, imagine exactly what will happen when the climber falls if you seat yourself in various positions. Begin with the direction of the fall in relation to yourself seated in each of these positions. Figure 29 shows a satisfactory position in relation to

Figure 29
Direct top belay.

The stability of the belayer's position—the sitting hip belay—is obvious. The belayer's right hand is the braking hand and his left hand is the guide hand. Note also the tautness of the rope between belayer and the stalagmite to which he is anchored. The climber's position directly below the belayer gives this type of belay its name.

direction. Figure 30 shows a poorly positioned belay in relation to direction. It also shows several other belay undesirables. It is obvious, in Figure 30, that the climber will pendulum and could crash into the rock face as he swings. In this situation the additional rope required

Figure 30
Poorly positioned belay.

The instability of the belayer's seating is obvious. Other dangerous errors can be seen. The belayer is inattentive and has allowed too much slack to develop between the climber and himself so that the fall would be a long one. Because of his unstable seating it would be difficult for him to arrest a long fall, and he would probably topple sideways. In addition to this, the belayer's anchor is so slender that the stress placed on it by a fall might well snap it, and there is so much slack between the anchor and the belayer that the climber's fall would drag him over the edge. The impact of this long fall would sever the rope at the unpadded ledge.

by the diagonal means that the climber will fall that much farther before the slack is taken up and his fall arrested. This places more shock load on the rope than is absolutely necessary and makes it more likely that the climber will be injured by the belay rope around his body when his fall is arrested.

After the result of the direction of the fall is examined, the effect of the fall on the belayer has to be considered. It is apparent that the impact of the fall will pull the belayer out of his spot. To counteract this, a way of sitting that will provide the greatest stability for the belayer is necessary. Figure 30 shows an obviously unstable seating and Figure 29 shows a good one, the sitting hip belay. This type relies on its tripod configuration for its stability. Despite this inherent stability it must be combined with an unquestionably secure anchor to give any real safety to the belay. If a cave formation is used as an anchor it must be of sufficient girth to make brittle fracture impossible. Don't use formations that are not firmly cemented to a rock base. Formations that have grown on a mud base, even large formations, have little strength. For extra security, a second anchor point can be utilized. A detail of anchoring a belay that is frequently overlooked is the amount of slack which should be allowed between the anchor and the belayer. There should be none. It is easy to imagine what would happen if the belayer is sitting two feet from the edge and there is three feet· of slack between him and the belay anchor. If the climber falls, the impact will drag the belayer over the edge. The same thing could happen if the belay rope is very elastic, the belay is set up too close to the edge, and the climber falls.

A second type of belay often used in caving is the standing belay (see Figure 31). This is used to best advantage when the climber is moving upward from the belayer. One important safety reminder is that the belayer should not stand directly under the drop to operate this type of belay. Falling rocks are always a hazard, even when the caver is wearing a hard hat.

The discussion thus far has pertained to uncomplicated belaying situations where the climbs encountered after entering the cave are initially downclimbed. In these, it is usually possible to eliminate all of the undesirables (see Figure 30). However, there are at least two situations in cave belaying where the preceding information is not adequate by itself and where the undesirables can't be eliminated but must be compensated for.

The first of these situations occurs in caves where the climbs encountered after entering the cave initially have to be ascended. In this case, the most adept climber will lead the climb belayed from the

Figure 31
Direct bottom belay with double rope rigging.
Note that the belayer is anchored to a stalagmite. His right hand is
the guide hand and his unseen left hand is the braking hand. The
belay rope passes under the belayer's buttocks.

bottom. Unlike the bottom belay previously mentioned (Figure 31), there is no double rope bottom belay rigging (see Figure 28) set up at the top to act as an intermediate point of support between himself and the belayer in case of a fall. He must carry and set up supports (pitons, carabiners, and slings) as he climbs. Every few feet, as any suitable rock projection appears, he braces himself as best he can, hooks the sling with carabiner attached over the projection, threads his belay rope through the carabiner, and continues to the next likely projection (see Figure 32). Sometimes, projections are scarce or absent and the climber has to make do with cracks suitable for the insertion of one of many types of piton. After hammering the piton into the crack, he hooks a carabiner to its end and threads his rope through the carabiner.

Figure 32
Running belay for indirect belaying.
Note the intermediate points of support (piton and carabiner, sling and carabiner) between the belayer on the left (unseen) and the climber.

Because, in such lead climbing, the climber is usually moving above or across from his last intermediate point of support, he could fall many feet before he is caught. It should be noted that this differs from the situation of the climber in Figure 31 who is not vulnerable to a long fall because his intermediate point of support is directly above him. Although both types of belay have an intermediate point of support in common and this makes both "Indirect Belays" in mountaineering language, only the former could be called a "complex" belay, using the vulnerability of the climber as a measure of complexity.

A traverse is a second situation in which the undesirables can't be completely eliminated. A traverse involves crossing a drop rather than descending or ascending it. Because the climber is moving *across* from his belayer and intermediate points of support, he is continually exposed to long falls. The lead climber traverses the drop, belayed from behind, and sets up pitons or slings for himself and the rest of the party.

It should be apparent from the discussion on complex belaying that since very little protection is available to the climber, he must therefore be an expert. The belayer, too, must be an expert. He not only has the difficult task of stopping a long fall but he has to do so while minimizing damage to the climber and to the rope. Lastly, if pitons or other hardware are used, the techniques involved need to have been *mastered* by the user before any protection can be gained from them. Trusting a life to a poorly placed piton is potentially lethal. Safely coping with complex belaying, therefore, demands advanced techniques that are beyond the scope of this book. A climbing club or school is the best place to go for necessary instruction. However, a good climbing book will be a *help*. Practice what you are taught outside the cave, preferably in limestone, which is the rock on which the techniques will ultimately be used.

Stopping the Fall

There are two ways of arresting a fall: static belaying and dynamic belaying. Static belaying is the method of most use to cavers. Dynamic belaying need only be used where a long fall is possible. At one time it was considered good practice to apply a dynamic belay to all falls, but there is now strong feeling that a good belay rope will give an adequate dynamic quality to even a static belay because of the elasticity built into the rope for that purpose. A leather glove must be worn on the *braking* hand while belaying. To arrest and hold a fallen caver is quite painful. You should be prepared for this and not let the surprise of the unpleasantness cause you to let go. The pain of the rope burns on the braking hand would be unbearable if the fall occurred when the hand was ungloved. The *guide* hand, however, can be gloved or ungloved according to personal preference.

Static Belaying

Static belaying is little more than the *immediate* arrest of a fall. The rope from the climber comes up between the legs of the belayer to the guide hand, passes around his lower back and hips, then is held by the braking hand (Figure 30), which should never leave the rope. In the event of a fall, the braking hand immediately grasps the rope and the hand is moved across the abdomen. The large friction area provided by the back and abdomen makes it possible for the belayer to hold a body weight and, if necessary, lower the caver to safety. Always make sure when setting up the belay that there will be enough rope to lower the caver to the ground or to an intermediate safe place if this should become necessary. It is almost impossible to haul someone *up* to safety unless the belayer has the help of extra manpower. It is important not to attempt to arrest the fall with the guide hand. If the guide hand is tightened, the friction areas of the body are not brought into use, and the belayer is left holding the whole weight of the climber without any assistance from the friction areas. The best place for the guide hand when a fall is arrested is

Figure 33
Pulley and weight system for belay practice. This pulley system will provide a means of learning belaying techniques without risking lives. Belayer supports weight of 150-lb. sack of wet sand. At intervals he lets go of the rope, allows the sack to drop for a few feet, then arrests the fall. Use an old or rejected rope for this exercise. Otherwise, it will ruin a good rope for belaying.

by your side with palm on the ground. In this position your hand and arm are strengthening the stability of your seating. The static belay can be used safely on climbs where it is unlikely that the climber will have fallen more than about 3 to 5 feet before the arrest is effected.

Dynamic Belaying

Most of the features of the static belay also hold true for the dynamic belay. The greatest difference is the immediacy of the arrest and the distribution of the shock of the fall. On a long fall, if the stop is too sudden the rope may reach its breaking point, and the shock to the climber's body may cause serious, perhaps fatal, internal and injuries. Therefore, it is desirable to slow the fall over a few feet of rope before bringing the caver's fall to a complete halt.

Belaying the Cable Ladder Climber

Generally, there should be some tension on the rope between the climber and the belayer. When down-climbing, the rope should not be paid out. The downward movement of the climber should drag the rope through the braking hand of the belayer, with just enough tension being applied to prevent slack rope from developing. When belaying the climb back up, the rope will have to be taken in quickly and smoothly by the method shown in Figure 34. The braking hand should never come off the rope. If an exhausted or injured climber is being belayed, considerable help can be given to him if the belayer takes some of his weight with the rope as he climbs.

Belaying Rock-Climbing

Rock-climbing belaying is much the same as belaying cable ladder climbing. One difference, however, is that the need for slack and tension varies from minute to minute. The climber determines how much of either he needs at any given moment and relays the information to the belayer. The belayer must always remain alert to these requests.

Climbing Signals

There is a large variety of climbing signals applied in caving. Those that are chosen usually depend on the caving conditions. If there is falling water nearby and/or the drop is a long one, verbal signals may be useless. Whistle signals are the most useful in this case. Verbal signals should be kept simple. Requests for "tension," "slack," "up

Step 1. Braking hand (left hand) directs the slack that is taken in from the climber with the right hand, forward, toward the coil of rope.

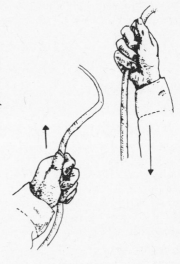

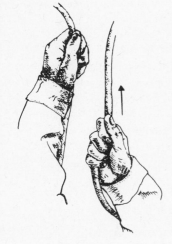

Step 2. Braking hand grips the rope and the right hand slides forward along the rope.

Step 3. Right hand grips both ropes and the braking hand slides back toward the body to repeat Step 1.

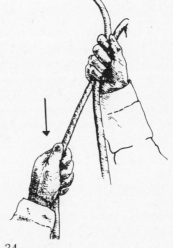

Figure 34
Taking in rope while belaying.

rope," and "stop" are easily conveyed and understood. The particular method of conveying information should be agreed upon *in advance* by all members of the group. Whistle signals are fairly standard. The most commonly used set follows:

One blast—STOP
Two blasts—UP
Three blasts—DOWN

Belaying the Rappelling Caver

Top belaying a rappelling caver will often cause problems in a cave, especially in free-fall rappels where belay rope and rappel rope can twist about each other and immobilize the caver. It is a good idea to learn how to rappel with a top belay, but this learning should take place outside the cave.

The bottom rappel belay is often used in cave rappelling as an alternative. This method does not require a regular belay setup. It is effected by one person standing at the bottom of the drop. His task is to pull hard on the rope if the rappeller has lost control or to apply some tension to the rope if the rappeller has too little control. It can be 100 per cent effective if the arrest is almost immediate, but appears to be less effective and has been known to fail completely if the rappeller has fallen any distance and gathered speed before the attempt is made to arrest his fall.

Bottom rappel belaying has one serious disadvantage which must always be taken into account when deciding whether or not it should be used. To be certain that he will be able to arrest the fall, the belayer should be as directly beneath the rappeller as possible, and there should be no slack in the rope between his arrest hand and the rappeller. But this optimum position for belaying puts the belayer in the most vulnerable position possible. He is directly exposed to any rocks kicked down by the rappeller or by people at the top of the pit. Bottom rappel belaying gives safety to the rappeller at the expense of the belayer. If it must be used, the safe way is to find an overhang under which the belayer can stand, not too far from the line of descent. But the effectiveness of the belayer is reduced as he moves farther from the line of descent. The problem of falling rock besets the double-rope bottom belay also. However, being directly under the line of descent is relatively unimportant for an effective belay in this case, so one can move far away to gain protection under an overhang.

Whether or not to bottom belay a rappel is best determined by the pit you are exploring. If it can be done safely—*i.e.*, with the belayer well protected and still able to belay effectively—it can and should be done. In other circumstances it has dubious value as a safety measure. Statistically, falling rocks have caused more injuries than have rappels (excluding those accidents that were not attributable to loss of control: *e.g.*, broken seat slings, broken rappel rope, faulty equipment, etc.). Like all other vertical techniques, it should be practiced outside the cave.

In the cases where belaying a rappel has been decided against, the only protection available is adequate training and practice for that particular drop, very careful rigging of vertical equipment, and caution while rappelling. All these add up to a competent self-belay.

Rock-climbing and belaying, beyond the level presented in this book, should be learned from mountaineering experts. There are a number of excellent climbing schools in the United States, several of which are listed in the Recreational Equipment, Inc. catalog and in *Summit* magazine. If climbing school is beyond your budget, join a climbing club for a while.

8.   Rappelling

Rappelling is a means of descending a pit on a fixed rope utilizing the friction caused by wrapping the rope around the body, by rappel devices, or a combination of both, to control the rate of descent. Such body wrap rappels, carabiner wrap rappels and their combination were originally developed as mountaineering techniques. Over the last 15 years, however, U.S. cavers have used these techniques as the basis for the design of rappel devices which are suitable for pit caving. From their innovations emerged the now popular tandem brake bar rig (see Figure 35) and the rappel rack. While both of these devices

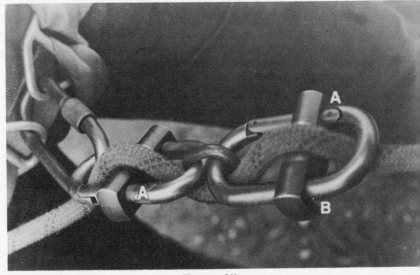

Figure 35
Correct configuration of tandem brake bar rig.
Note that the carabiner gate opening (A) is uppermost—away from the caver—and the rope passes over the side of the brake bar that is opposite to the brake bar opening groove (B).

are very useful, the tandem brake bar rig is most suitable for smaller pits; ideally the province of all beginning pit cavers. As this book has been designed for the less-than-expert, particulars on rappelling technique have been confined to the use of the tandem brake bar rig.

Construction of Rappelling Equipment

A useful and inexpensive seat sling (see Figure 36) can be constructed from 1-inch tubular nylon webbing or 2-inch automotive safety web-

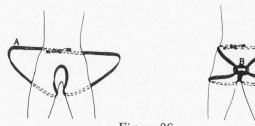

Figure 36
Seat sling for rappelling.
A—continuous loop of tubular nylon   B—locking carabiner

bing. Approximately 8 feet of webbing are joined with an overhand knot. To adjust it to your own measurements, first sit down and then put the sling on with the carabiner and the overhand knot in front of you. Draw the sliding end of the nylon through the knot and keep rearranging the loops around the legs and back and taking in the slack nylon until the sling fits snugly. Finish off the loop with a second overhand knot, using the sliding end of the nylon to form a Fisherman's Knot. A correctly adjusted seat sling is snug when you sit and uncomfortably tight when you stand straight. A locking carabiner completes the seat sling. Nonlocking carabiners in this position have been known to open during a rappel with disastrous results.

For a tandem brake bar rig you need two aluminum nonlocking carabiners, one strong chain link, and two brake bars, arranged in the configuration shown in Figure 35.

Rappelling Technique

Stand with your back to the drop, not too close to the edge, and lift the rope over your shoulder. Keep it there until you are completely rigged in. Pay special attention to putting rope into the rig so that the rope passes over the back of the brake bars on the opposite side to the brake bar opening grooves. The pressure of the rope is then closing the brake bars tightly on the carabiners. It is probably unnecessary to point out that putting the rope over the opening groove side of the brake bars will cause them to pop open and release the rope as soon as you begin your rappel. Note in Figure 35 that the openings of the carabiner gates lie uppermost. When rappelling, always be aware that loosely fitting brake bars can slip down toward the carabiner gate and distort the carabiner enough to allow the gate to open if its opening is below the brake bar. Choosing brake bars that fit firmly on the carabiner is a wise precaution.

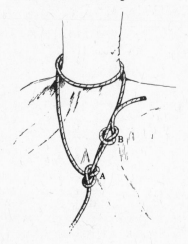

Figure 37
Fixed rope rigging for
rappel and Prusik.
The same criteria used for select-
ing a rigging point for a belay
also apply to rigging a fixed rope
for rappelling and Prusiking.
A—single bowline
B—overhand knot

If there is any slack rope between the tie-off point and your brake bars, take it in. In almost every situation, walking backward over the top of the drop is the smoothest and least abrasive means of beginning a rappel. Walk backward to the edge, then spread your feet about 2½ feet apart. Then, keeping your feet stationary, let out a little more rope and lower your backside until your legs are at an angle of about 45 degrees. This leg angle and widely planted feet provide a very stable and therefore safe configuration while rappelling against rock. The wide planted feet give excellent lateral balance and stability. The climber who starts out with feet too close together may well lose his balance, swing to the side and crash into the rock with a shoulder. This can be a particularly unpleasant experience. Keeping the feet in front of the body rather than below makes it possible to rappel down a sheer wall because the whole surface of the sole of the boot can be used for traction (see Figure 38).

Figure 38
Correct rappel stance.

Beginning the rappel by walking backward over the cliff is not possible if the pit entrance is very small or if the cliff is undercut right at the lip. In these situations the best thing to do is to lie on your stomach and lower your legs over the cliff. Make sure that clothing, hair, etc., does not become entangled in the brake bars as you slide over. Lower your body until you are right over the edge.

To change from a wall rappel to a free-fall rappel in which you will be in sitting position, at the point where there is no more wall on which you can put your feet, stop, extend the nonbraking hand to the wall, then lower one leg at a time until you are in a sitting position. Lower yourself still further, then withdraw your hand after your head is clear of the overhang.

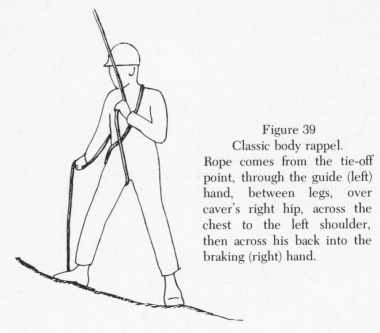

Figure 39
Classic body rappel.
Rope comes from the tie-off
point, through the guide (left)
hand, between legs, over
caver's right hip, across the
chest to the left shoulder,
then across his back into the
braking (right) hand.

The braking hand should be about mid-femur, about 10 to 12 inches
from the bottom of the brake bars (see Figure 38). Many different
degrees of braking power of friction will be experienced while rappelling. With the greatest amount of friction—generally felt at the top
of a drop as a direct braking effect of the weight of the rope below—you
may have to feed the rope into the brake bars in order to move.
The next stage is the most pleasant, a controlled rappel. The last
stage is that of decreasing control. The last stage causes many
accidents. However, simple precautions during the rappel can circum-
vent most of these accidents.

Rappel slowly and smoothly. Don't be tempted to zip down the
rope. Five minutes for a 100-foot rappel is a reasonable speed. A
moderately slow rappel allows some of the heat that builds up in the
brake bars to dissipate, and also gives you a little time to think about
what you are doing. On the other hand, a fast rappel can build up
enough heat to melt rope. When normal control is experienced, the
hand should be about 10 to 12 inches below the brake bars and held
out from the leg. When you need more control bring the hand hard
against the thigh. As more is needed, the 10 inches is lengthened
to about 14 inches and the hand is moved behind the buttock with

the rope tight against the body. To increase friction further, the rope can be guided into wraps around the leg. Generally, one or two wraps is sufficient for any rappel, but a new slick rope might make more necessary. If you need more or need to stop completely, take the leg wraps off. Then manipulate the rope with the feet until it passes under the instep and over the same foot, then straight under and over the other foot, then jam one foot hard and on top of the other foot. Stand on your foot wraps.

A new, slick rope is going to provide a faster rappel than an old nylon rope which has developed surface fuzz and has become stiff. Several changes in friction will be felt while rappelling. These changes are caused by a combination of several factors—two of the most important being the initial braking capacity of the rappelling device and the weight of the rope below the caver.

Special Cautions

1. If you feel a need to hold onto something with your nonbraking hand, for security or balance as you back over the edge, hold the brake bar rig, don't hold the rope. Holding the rope above the brake

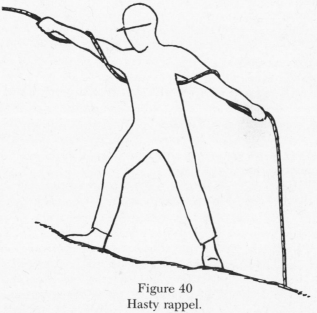

Figure 40
Hasty rappel.
Note that the body is at right angles to the cliff face.

bar rig is acceptable if it is held lightly so that it is free to slide through your hand as you descend. Many novices have a tendency, however, to clutch the rope tightly above the brake bars because of panic feelings. This, in effect, leaves them supporting their whole weight with their nonbraking hand, without the help of the brake bar rig. You can clutch the brake bar rig very tightly without losing any braking power.

2. If you have long hair, tie it up underneath your hat. Hanging long hair usually gets caught in brake bars and this is an excruciating experience. Loose clothing, hair ribbons, scarves, beads, etc., also get caught up. This has occasioned several rescues.

3. Always keep overhangs in mind when you are on the way down. It is possible to swing back in toward the overhang at head level after your feet clear it. To prevent the possibility of whacking your head, keep your nonbraking hand extended in front of you until your head is clear of the overhangs.

4. Under no circumstances let go of the rope with your braking hand. By mischance you could possibly flip upside down while rappelling. This experience will be unpleasant but not fatal if you don't panic. All the while that you have control with your braking hand, even if you are upside down, you can continue your rappel. Don't let panic make you let go with your braking hand.

5. Most of the techniques outlined above need much practicing. They should be polished until they are automatic responses and this initial polishing should not take place in a cave.

6. Don't put off learning these techniques under the mistaken belief that you can be top belayed or bottom belayed in the cave by someone else. The only safe belay for cave rappelling is a competent self-belay.

7. Shock loading such as a sudden stop after a fast or uncontrolled rappel is particularly dangerous on tandem brake bar rigs. It may even result in gross distortion of the carabiners holding the brake bars. If the distortion is sufficiently gross, the rope can pop right out. This is good argument for a smooth, carefully controlled rappel.

8. While it is well to know both the Classic Body Rappel and the Hasty Rappel in the event of an emergency, they should be avoided in caves unless absolutely necessary. (See Figures 39 and 40.) Your first few tries should be belayed from the top and should be against a rock wall. The Body Rappel can be used on a vertical wall but a Hasty Rappel can be used only on a sloping wall. Use neither technique on a drop of more than about 30 feet. On a free-fall rappel, never use a Hasty Rappel and don't Body Rappel if the free fall is more than 10 feet in depth.

9. Carabiner Wrap Rappels have a high accident rate in caving. Avoid this technique.

10. If there is a possibility that the bottom of your rope is not long enough for the pit and has not touched the floor, tie a big knot in the end of the rope, then rappel down with your Prusik chest harness on you and your Prusik knots at an accessible point at your waist. While you are set up this way, it is not likely to be a traumatic experience if the end of the rope appears when you are still 20 feet or so from the floor. Simply immobilize yourself, put on your chest knot and two foot knots above your tandem brake bars on the rope. Clip your chest carabiner into the chest Prusik, hang from the chest knot, release your feet from the rope lock, then put each of your feet into a foot sling. Detach your brake bars from the rope and then Prusik up.

Descending Deep Pits

The depth at which a pit is considered deep will vary from one caver to another. However, a useful definition of a deep pit is one so deep that the tandem brake bar rig cannot reliably control the rate of descent. For safety's sake, this could be considered to be about 150 feet. If deep pit caving is on your agenda, postpone the event until you have built up experience in many smaller pits. After that, the next step is to buy a rappel rack. This item, a U-shaped steel rod with six aluminum brake bars bridging the "U," is a necessity for a long rappel. Practice manipulating the brake bars to vary the amount of control and to immobilize yourself on an outside cliff or small building, then find cavers who regularly explore deep pits. Your first few deep pits should be done with them. If your group doesn't number such people in its membership, write to the National Speleological Society for the addresses of Grottos whose members do deep pit caving as a regular activity.

9.   Prusiking

Like rappelling, Prusiking is a technique which originated in moun-taineering, where it is used for crevasse ex.t. Prusiking describes any means of ascending a fixed rope with friction knots or mechanical gripping devices, and provides a logical means of getting out of a pit after rappelling into it. Cavers have been devising varieties of friction knots and mechanical ascenders for many years, and today a caver has a choice of many ways to ascend. The Prusik knot (see Figure 41) is the best-known friction knot and, compared to the other friction knots, is the most reliable, most versatile, most easily recog-nized and the simplest one to tie. These reasons explain its detailed presentation in this book as the best choice for ascending a small pit even though it is a strenuous pastime and a somewhat unpleasant means to the end of getting out of the pit. Because of this, it is seldom practiced with as much dedication as it deserves. Both the fatigue it produces and the unpleasantness can be reduced by doing it efficiently and well. To this end, the following points should be kept in mind:

- Make sure that your Prusik knots are adjusted to your length and that your chest harness is comfortable.
- Practice, practice and practice until you have developed an easy 1-2-3 rhythm. This rhythm is the best means of pacing yourself on a long climb. It requires a great deal of co-ordination, and this only comes with practice.
- Avoid stopping and starting. Starting requires a tremendous amount of energy. Aim at setting a constant, comfortable pace for the whole distance.

Prusik Technique

1. While standing on the ground, loosen, or "break," the chest knot until it is very loose; then move it up the rope as far as it will go.

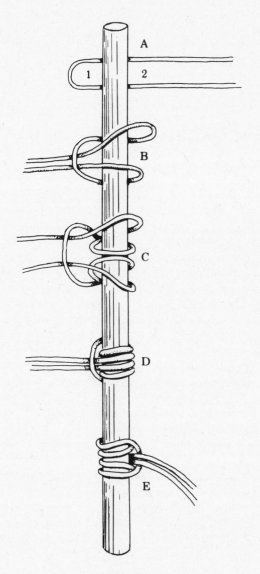

Figure 41
Construction of a right-hand Prusik knot.
With whole Prusik sling behind fixed line, make short end of loop (1)
on the left and the long ends (2) on the right as in (A). Loop long ends
through short end (B) . . . twice (C). Rearrange knot to look like (D)
and then pull knot tight. Rotate knot so that long ends protrude
from the right side to form a finished right-hand Prusik knot (E).

2. Flop down until body weight is all on chest harness. Keep hands away from chest knot when you flop down. Holding the knot prevents it from tightening and it will slip back down the rope. (See Figure 42.)

3. While suspended by the chest harness, thoroughly loosen the top foot knot and move it up the rope to about 1 inch below the chest knot. (See Figure 43.)

4. Thoroughly loosen bottom foot knot and move it up to 1 inch below the next knot. (See Figure 44.)

5. Thoroughly and quickly break chest knot as you begin to stand up in your foot knots. Move up the rope with a combination of pushing

Figure 42
Prusik technique.
Flop down until body weight is all on the chest harness.

Figure 43
Prusik technique.
Move top foot knot up to chest knot.

up with your legs and pulling up with your arms (see Figure 45). By breaking the chest knot completely *before* you have pulled yourself up, it is possible to move the chest knot up the rope with one hand at the same time that you are pulling up with the other hand and standing up. A fatigue-producing mistake that many people make at this point is to pull themselves up with one hand and, while supporting all their weight with that hand, then to break the chest knot with

Figure 44
Prusik technique.
Move bottom foot knot up to top foot knot.

the other hand and move it up the rope. The obvious undesirable
here is the 10 to 20 seconds during which most of the body weight
is supported by one arm. It is easy to imagine the disproportionate
strain placed on the pulling arm by such a technique. As far as possible
the legs should take most of the strain of moving you up the rope.
This is best accomplished by beginning your stand-up/push-up from
a sitting-on-the-heels position.

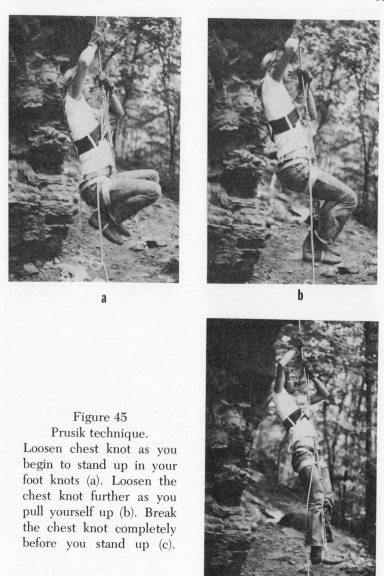

Figure 45
Prusik technique.
Loosen chest knot as you
begin to stand up in your
foot knots (a). Loosen the
chest knot further as you
pull yourself up (b). Break
the chest knot completely
before you stand up (c).

6. Flop down in the chest harness at the same time that you begin to break your top foot knot (see Figure 42).

Briefly, coordinating your movements to produce the rhythm necessitates the following:

Break chest knot and stand up—*One*
Break top foot knot and move it up—*Two*
Break bottom foot knot and move it up—*Three*
Break chest knot and stand up—*One*, etc.

The importance of completely loosening or breaking the knots (see Figure 46) cannot be too heavily emphasized. Not only does it require

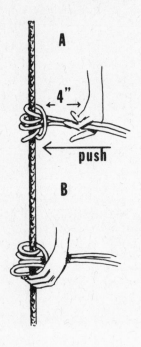

Figure 46
Loosening the Prusik knot.
A. First, push sling toward the knot until the knot is completely loosened.
B. Then, with two fingers supporting the loosened knot, move the knot up the rope as necessary.
This procedure is guaranteed to eliminate most of the wear and tear on the rope.

much less energy to move a loose knot up the rope rather than having to push a tight knot up but also it virtually eliminates wear on the Prusik knot and the main rope. On the other hand it is not unknown for a person who either ignores the loose knot edict or doesn't know about it, to wear right through a set of Tenstron knots in 100 feet of Prusiking.

Figure 47
Prusik technique.

Resting is sometimes necessary on a long climb. Also, the resting position is convenient for changing the chest knot for a new one if this becomes necessary. Break foot knots and move them up to about 8 inches below the chest knot. Sit on your heels and rock forward and embrace the rope. This takes the weight off the chest and is quite comfortable.

Prusik Difficulties

Slipping knots. As mentioned before, unconsciously holding the knot after your weight is on it will interfere with its locking properties and cause slipping. Muddy Prusik knots and/or muddy fixed rope can easily cause slipping. Using Tenstron or Polypropylene rope for Prusik knots will reduce the possibility of mud-caused slipping, but even these will allow slip under some conditions. In mild instances of slipping you may be able to rework the knots with your fingers to tighten them. In more severe cases the triple Prusik wrap or the tandem Prusik knot can be used (see Figure 48). These will slow down your Prusik, but at least will get you out of the pit.

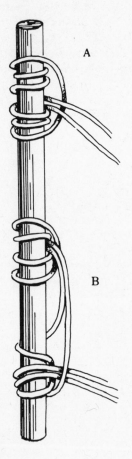

Figure 48
Prusik technique.
A. Triple Prusik wrap.
B. Tandem Prusik knot.
Either of these will slow down your Prusik, but will get you out of the pit.

There are several ways of reducing the amount of mud on the rope. One way is to carry the rope through the cave to the drop in a duffle bag. In this way the duffle bag gets coated with mud and not the rope. This method has another advantage in that the bag affords considerable protection to the rope from abrasion and snagging and will keep it in better condition. The same bag can be used as the always necessary rope pad. Don't use your rappelling and Prusiking gloves for caving. Muddy gloves don't much affect rappelling, but can turn a clean rope into a greasy barber's pole. This can make climbing-out a miserable experience for the last person up the rope.

Binding knots. Prusik knots that bind up and won't move can be a great menace and have caused at least one known fatality. It is sometimes possible to loosen them with a great deal of effort. However, as in so many other things, prevention is the best cure. There are certain ropes, and combinations of ropes and circumstances, that will almost ensure that binding will happen. Nylon Prusik slings on nylon rope—*i.e.*, Goldline slings on either Goldline rope or Bluewater II rope—can cause binding if both the slings and rope have a fine surface fuzz from wear. If either or both slings and rope are wet, binding will probably occur with nylon on nylon. The sensible choice of Prusik sling material (see page 88) is the best preventive measure.

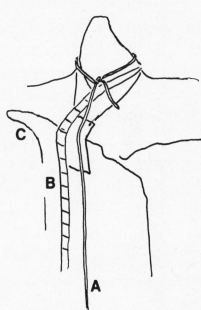

Figure 49
Prusik technique.
Severe overhang (C) is handled with cable ladder (B) in addition to fixed line (A).

Overhangs. Prusiking out of a pit with a sharply undercut overhang at the lip of the pit can be a nightmare if a few simple precautions are not taken when rigging the pit initially. The most desirable of these is the rigging of a short length of cable ladder from the tie-off point so that about ten feet is hanging over the edge of the drop alongside the rope (see Figure 49). The climber can then transfer his weight onto the rungs of the ladder and lift the rope off the lip so that the chest knot can be pushed up and over. This process can be repeated until all the knots are over the lip. *Don't* remove your Prusik knots and continue the climb out of the pit on the cable ladder. Stay on the rope. The ladder is as unsafe as any unbelayed ladder climb. The second precaution is more difficult to accomplish, but is still a useful technique. Rig the pit with the rope so that the short end of the rope goes back over the edge and hangs about ten feet below the lip (see Figure 50). When coming up, the climber should move all three knots right under the lip, break the chest knot, transfer

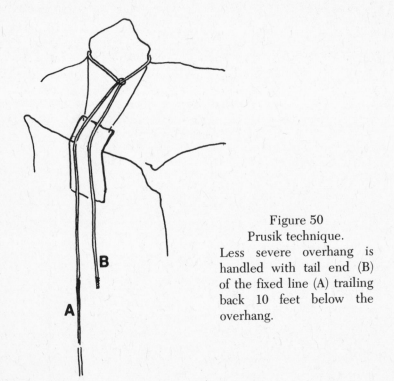

Figure 50
Prusik technique.
Less severe overhang is handled with tail end (B) of the fixed line (A) trailing back 10 feet below the overhang.

his weight to the short end of rope and quickly slip the chest knot over the lip. This is repeated with the other knots. A less extreme overhang can be negotiated by moving all the knots under the lip, breaking the chest knot, pushing out from the wall with the feet and in the brief moment that this lifts the weight off the rope, the chest knot can be pushed over the lip.

NOTE: *Do not remove your Prusik knots until you are in a safe place away from the lip.*

Putting Knots on the Rope

For ease of handling, knots should be put on the rope with right-hand knots alternating with left-hand knots. If you are right-handed and therefore strongest in the right arm, you will be pulling up with your right hand; so the chest knot will be manipulated with the left hand and should be a left-hand knot. The top foot knot will be moved with the right hand and will be a right-hand knot, and the bottom foot knot will be moved with the left hand and will be a left-hand knot (see Figure 51).

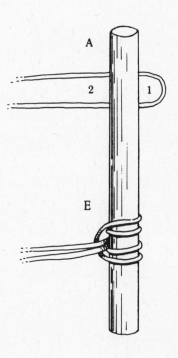

Figure 51
Construction of a
left-hand Prusik knot.
With the whole Prusik sling be-hind fixed line, make short end of loop (1) on the right and the long ends (2) on the left as in (A). Complete the knot as shown in steps B, C, and D of Figure 41 to form a finished left-hand Prusik knot as in (E).

Choice of a Chest Harness

For comfort, a horse girth is unbeatable. It has to be loose fitting. A 36-inch girth is loose enough for all but very large people. It should be backed up with a continuous loop of 1-inch tubular nylon, independently suspended from the same chest carabiner. Although I know of no instance where a horse girth has actually broken while being used by a caver, little is known about its true strength. For this reason, a backup system of tubular nylon around the chest provides a margin of safety. The nylon and the chest carabiners by which it is held have a known and more than adequate breaking strength. See Figure 52 for details of chest harness construction.

For the leaner pocket a horse girth can be constructed with 100 feet of parachute cord and two saddlery rings. Although this type of girth is not quite as comfortable as the commercial variety, it is inexpensive and can be custom made to fit anyone whose chest is so large that the regular 36-inch size is not satisfactory. Borrow a commercially made girth and copy the construction. It is important that the safety loop of tubular nylon be used with the homemade variety as well as with the commercial variety of horse girth.

For very short climbs the harness shown in Figure 53 is used by many cavers. However, it usually becomes uncomfortable on a climb of over 60 feet.

Prusik Knot Materials

The most desirable material currently available for Prusik knots is ¼-inch Tenstron. Second and third choices are 5/16-inch Tenstron and 5/16-inch Polypropylene. Both Tenstron and Polypropylene are trade names for rope made of polypropylene fiber. Tenstron is a brown-colored laid rope and Polypropylene is a bright yellow, round, tightly braided rope. Do not confuse it with the yellow, loosely braided polypropylene fiber rope that can be bought in most boating and Army surplus stores. The Tenstron is a laid, hard-fiber rope that is quite pliable and therefore easy to manipulate. It functions extremely well under wet and muddy conditions. The least desirable is manila rope. This is a natural fiber which will rot under usual cave conditions. In addition, its limpness causes Prusik knots to slip. Next in line as least desirable is Goldline. When very new, Goldline functions reasonably well, but with even slight usage it becomes hard, unpliable, and covered with a fine surface fuzz which can make Prusiking a nightmare

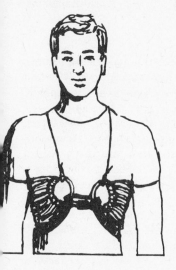
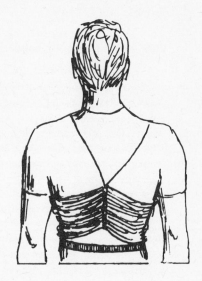

Figure 52
Horse girth used as Prusik harness.

A. 36-inch horse girth.
B. Parachute cord shoulder
   loops.

C. Locking carabiner.
D. 6-foot loop of 1-inch tubular
   nylon with ends joined together
   with a Fisherman's Knot.

Adjustment: The sling should be loose. A 36-inch girth with three locking carabiners should be loose enough for even big-chested people. A 36-inch girth with one locking carabiner may be loose enough for a very slight person. Don't buy a girth of less than 36 inches and don't wash the girth when it is dirty—it will shrink about 4 inches when wet.

Note: Hook chest Prusik sling into *both* carabiners if two are used.

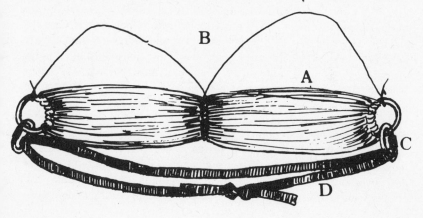

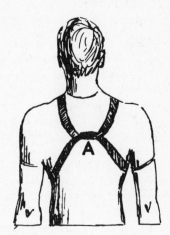

Figure 53
Simple Prusik sling for chest.

A. Twist once as shown.
B. Locking carabiner.
C. Arm loops.

D. Continuous loop of 1-inch tubular nylon joined with a Fisherman's Knot.

Adjustment: The harness should be tight, but not so tight that breathing is restricted.

when climbing other nylon rope. Both manila rope and Goldline are unreliable in water and mud.

Tenstron has one inherent disadvantage. Like other rope of similar material its fibers abrade and break relatively easily. However, this problem can be countered by: (1) becoming paranoid about thoroughly loosening knots before moving them up the rope (this almost eliminates wear on the knot), and (2) *always* checking knots after your climb. If any more than 1/8 of the diameter of the Tenstron is abraded, replace the worn knot or knots before you climb again. The chest knot wears faster than the foot knots, so it is a good idea to always carry a spare chest knot in an accessible place. In a situation where there is a coating of gritty mud on the rope it is possible that wear on the knot will be accelerated. It may then become necessary to replace a knot or knots while on the rope.

Preparing for Rappelling and Prusiking Predicaments

Much of safety in vertical caving lies in being prepared for adverse situations. Adequate preparation can make it highly improbable that any of the following mishaps will occur, but when even a slight possibility exists it is wiser and safer to know what to do to get yourself out of the predicament.

Chest knots can and do break, leaving you suspended upside down from the rope by your ankles. The rope may prove to be ten feet short of the bottom, which means that the first man down has to change over to Prusik harness and come back up without touching the ground. Braking devices can get overheated and cause a too fast rappel. Knots can jam or slip. All of these things can be quite traumatic if, on the first time they happen to you, you don't know what to do, especially if you are in a cave.

By rigging the rope in a gym or tree on the pulley system shown in Figure 54 it is possible to practice remedies for all these situations. Learn how to replace one complete set of Prusik knots with another complete set while on the rope. Deliberately detach your chest knot while you are on the rope. From the upside-down position that will result, haul yourself up to standing position, embrace the rope, put a spare chest knot on the rope, and then hook yourself into it. Change from rappel to Prusik rigs while on the rope. This requires the ability to immobilize yourself in rappel position so that both hands can be removed from the rope to begin to put on Prusik harness and slings. This immobilization technique was described in the section on rappelling. Practice tightening knots with your fingers and practice making tandem and triple Prusik knots to be ready for slipping knots. Soak some old Goldline or manila knots in water for several hours, then use the rope. It is quite an experience to Prusik with bound-up knots. Various leg wraps and/or adding or subtracting brake bars on a rappel rack to vary the amount of control you have while rappelling, along with ways to immobilize yourself in rappel position, should be practiced until they become automatic responses. Knowing what to do in a bad situation *and* how to do it reduces the trauma and makes panic unlikely. A cool head and precise reactions under these conditions will be your most valuable asset.

Prusik Sling Construction

It is important to make your own individual set of Prusik slings as soon as you are ready to take your Prusik skills into a cave. Construc-

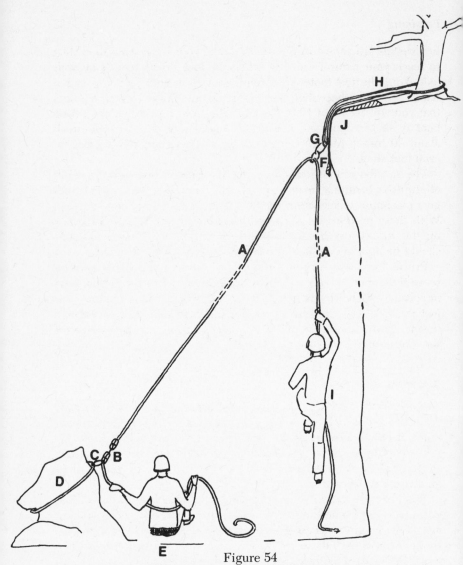

Figure 54

Pulley system for vertical caving practice.

A. Rope.
B. Tandem brake bar rig with . . .
C. Continuous loop of 5/16-inch Goldline or 1-inch-wide tubular nylon.
D. Anchor point for pulley system.
E. Belayer.
F. Rescue pulley with rope threaded through.

G. Carabiner link between pulley and loop H.
H. Continuous loop of sling material from overhead anchor point (overhead beam, tree limb, etc.) to carabiner.
I. Climber.
J. Rope pad.

tion details are shown in Figure 55. Make your longest finished sling to reach your natural waistline when the foot Prusik loop is on your foot. Your shortest foot sling should reach from your foot to your hip bone, and your chest sling (see Figure 56) should be of such a length that you will be about 10 inches from the fixed rope when the Prusik knot is on the rope and you are attached to it. Most people need about 30 inches of Prusik material for a chest sling and 15 feet for each foot sling.

The specific proportions chosen for the Prusik slings are the result of extensive testing by many experts, who found that these proportions give maximum efficiency in terms of speed and energy conservation. With these proportions you should be able to move 30 to 36 inches up the rope every time you stand up in your foot slings and you should be able to complete a 100-foot Prusik in 5 to 10 minutes.

Prusik loops are chosen for the loops around the boots because these tighten effectively when you stand in your foot slings. It is very important to have loops around your feet which will not allow your feet to slip out if your chest knot should break and you find yourself upside down. Such a situation is unpleasant but can be rectified if the loops around your feet hold firmly.

Ascending in Deep Pits

Mechanical ascenders (Jumars and Gibbs) are expensive to buy. They can be used in small pits, but offer very little advantage over knots at only a fraction of the cost. On a very long climb, however, it is possible to move very fast and easily on mechanical systems, so that once you think seriously about doing deep pits, they will be an attractive alternative.

It is a simple matter to buy ascenders. It is far less simple to find out how to set up and use them so that you get the best out of them and a full measure of safety as well. You can begin by reading *Prusiking* by Robert Thrun. Talk to cavers who regularly use ascenders in deep pits. The National Speleological Society's annual convention is a plentiful source of knowledgeable cavers and relevant technical sessions. In addition, you could direct specific questions to the N.S.S. Safety and Techniques Committee in care of the N.S.S. office in Huntsville.

After buying your equipment and amassing all the relevant information, you will notice much of it is inconsistent and confusing. This is your cue to find a place outside a cave where you can safely experiment. Rely on the considerable pit-caving experience you have gained in the past to evaluate all these systems and to experiment until you

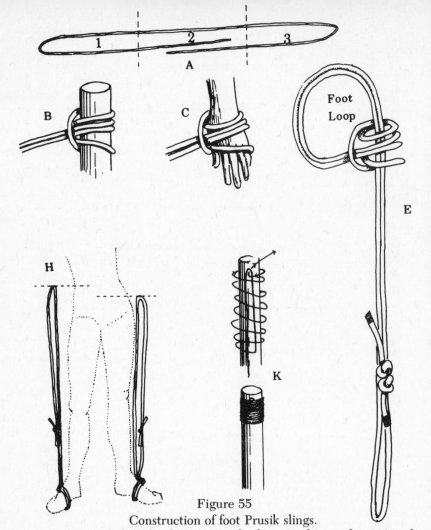

Figure 55

Construction of foot Prusik slings.

Cut a 120-inch length of Tenstrom for longest sling and a 110-inch length for the shortest sling and arrange each length as shown in (A). Tie section 1 into Prusik knot around broomstick as shown in (B). Slide knot off broomstick, enlarge it while being careful to keep loops intact. Insert hand to keep knot open, see (C). Pass sections 2 and 3 of sling material through Prusik knot to form foot loop, see (E). Put foot loops on your feet over your caving boots, tighten Prusik knot, then tighten foot loops. Tie half of the Fisherman's Knot (one overhand knot) in free ends of section 2. Raise the sling so that it reaches your hip bone (for short sling) and your waistline (for long sling), see (H). After adjusting each sling to your height in this manner, finish the Fisherman's Knot by tying the second overhand knot. Mark the long sling with a piece of string or colored tape to make it simple to identify in the cave. Whip cut ends of sling material with waxed carpet thread after melting and sealing them with a small flame, see (K). *Note*: The long foot sling will always be your *top* foot Prusik knot.

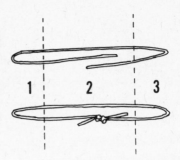

Figure 56
Chest Prusik sling.
Cut 30 inches of Tenstron and arrange as shown. Tie Fisherman's Knot in ends of section 2.

find one that is safe, efficient, and exactly suited to you. Don't take it underground until you are sure that all the bugs have been eliminated. A proper test of an ascending system is a climb on the pulley system, of at least twice the length of the pit in which you will be using it. Try to anticipate its performance in all the undesirables—mud, grit, water, ice, slopes, and overhangs.

10.   Cable Ladder Technique

Rappelling and Prusiking, together referred to as "vertical techniques," are not the only way to negotiate a pit; they are merely the most recently devised way. Much more time-honored is descent and ascent by cable ladder. The use of cable ladders in U.S. pits has become unusual, however, and this is a pity because they are a practical, although unexciting, solution to the negotiation of pits of any depth. They are particularly accommodating to the mixture of skill and ineptitude found among the participants on the average caving trip. It has become increasingly difficult to buy cable ladders in the United States. Those that are available are imported and expensive. However, it is feasible and relatively inexpensive to make them, and this is well worth the effort involved for groups that place a high value on simple and uneventful pit-caving trips.

Cable Ladder Construction

The lightweight cable ladder (see Figure 57) offers a simple way to ascend and descend a pit. The ready-made, store-bought variety (usually made in France) is expensive (about one dollar per foot). Homemade, the cost can be reduced to about 10 to 40 cents per foot. Such ladders are not difficult to make with the right tools, and a wealth of detailed information is available on materials and methods of construction and testing (see Reading List in back of book). Because your life will depend on your product, take no short cuts in construction or testing. Your initial ladder-making sessions should be supervised by an engineer or someone with comparable skills, and your ladders should be properly tested before they are used to make sure they conform to minimum safety standards. Also, don't compromise on quality of materials.

Use 12-inch rung-spacing in your ladders if you make them or, if you buy them. buy only those with 12-inch rung-spacing. This spacing lends itself to an easy rhythmic climbing style because 12 inches is a natural-sized step.

When you construct your ladders, spend some additional time making cable riggings for them. Several assorted lengths (for example, 15-, 25-, 30-, and 50-foot lengths) are very useful. Figure 57 illustrates the construction of these riggings.

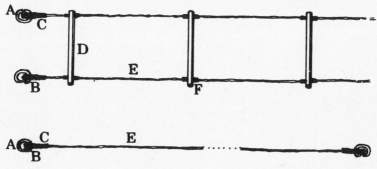

Figure 57

A. C link.
B. Thimble.
C. Crimped ferrules.
D. Aluminum alloy tubing rung.

E. Galvanized steel cables.
F. Crimped ferrule on either side of the rung for rung-fixing and spacing.

Climbing the Free-Hanging Ladder

Figures 58 and 59 illustrate two climbing styles for cable ladders. The first style (shown in Figure 58) is the *wrong* way. Climbing the front of a free-fall ladder places a tremendous strain on the arms and is exhausting after only 10 or 15 feet. Climbing on the front is acceptable only when the ladder is resting against sloping rock. Figure 59 shows the proper way—the classic heel-toe method. Both arms are around to the back of the ladder. One leg is around to the back of the ladder with the heel placed in the rung. The other leg is at the front of the ladder and the toe half of the boot is used to climb with. The body should be close to, and parallel with, the ladder. Climbing should be steady, balanced, relaxed and rhythmic. This style of climbing places the body on the ladder where it can most efficiently maintain a rhythmic climb.

Climbing is essentially a two-step operation. You move your *left* foot at the same moment that your *right* hand is moved to the nose-level rung. This is Step One. Then you move your *right* foot and *left* hand

Figure 58
Caver is climbing unbelayed and incorrectly.

Figure 59
Correct ladder-
climbing technique.

up one rung. This is Step Two. Repeat the sequence. Step up with your legs. Don't pull yourself up with your hands. The legs are much stronger and more adapted to this kind of activity than are your arms. The hands and arms should be used only for balance. A relaxed, rhythmic movement with steady breathing is the least tiring method of climbing. To help develop this, evenly count "1—2" aloud, in time to your steps. Avoid stopping and starting, because this uses up a disproportionate amount of energy.

When you are proficient you should be able to climb over 100 feet without needing to rest. An ascent time of less than 40 seconds for 100 feet, with a descent time of less than one minute, should be possible if your climbing is well coordinated. However, the most important features of the expert climb are steadiness and balance—not speed—so don't sacrifice these qualities in favor of speed.

Believe it or not, climbing down a ladder is more difficult than climbing up. Beginners often experience difficulty in locating the lowermost rung with each down step. Climbing by Style 2 makes rung location a little simpler, but the best aid is mentally to locate the elusive rung. After all, it is 12 inches *directly underneath* your other foot. It soon becomes quite a simple habit to think "12 inches" each time you take a down step. The other significant factor at this time is to keep calm and think clearly. Panic feelings tend to make that bottom rung only the more elusive.

Wear thick-soled boots when using a cable ladder. Otherwise, the rungs will cut into your feet if you are wearing soft-soled shoes. Most people find that placing the arch of their foot on the rung is the most convenient foot placement. A note of caution here: Don't climb a cable ladder while wearing boots with hook-type lacings. The hooks will catch on the cable and are almost impossible to extricate.

Climbing a Cable Ladder Resting Against Rock

If the ladder is resting against rock that is sloping rather than vertical, you will have to climb in front of the ladder and you may experience great difficulty in getting your fingers and feet onto the rungs. When the climb is particularly difficult—as when ascending over a smooth, rounded pit top—swing yourself and the ladder 90 degrees to the side (see Figure 60). In this position, the ladder will be at right angles to the rock with accessible rungs and you will be climbing with your shoulder against the rock. This is not a difficult technique, but it is tiring and should be used only when really necessary.

Figure 60
Climbing ladder at right angles to the rock.

In less troublesome spots, drag the ladder out with your hands until it is far enough away from the wall for you to be able to get your feet into the rungs. The toe of your boot will then keep the ladder clear of the rock so that you can get your hands on the rungs. When climbing against rocks, grasp the rungs at chest level rather than at nose level (see Figure 61).

On a drop that is very sloping and short, the ladder can be used for handholds while you walk down or up the rock. This technique is tiring to the arms, so it should not be used on drops longer than 20 feet. This is also a useful method for assaulting steep mud slopes.

It may become necessary to stop for a few minutes to adjust your equipment, to take a photograph, or to rest. At these times the ever-handy carabiner can be detached from your belt with one hand, clipped onto the belay loop around your middle and then onto a ladder rung at about the level of your chest. While attached in this way to the ladder, both of your hands are free to accomplish the task. *Note:* This is a comfortable resting position.

Miscellaneous Difficulties

When lowering a belay rope down next to the ladder it often happens that the rope will thread through the ladder. This causes a problem to the climber on his way up. He has to stop, untie his belay, unthread the rope, then retie his belay (all with one hand). Apart from the inconvenience involved, it also is dangerous. This contretemps can be avoided by coiling the rope loosely at the top of the drop and throwing it out over the edge, well away from the ladder. An alternative procedure is to lower the rope with a cave pack tied to the bottom.

If, however, you find yourself confronting the belay rope threaded through the ladder while on your climb, attach yourself to the ladder before you begin to untie your belay. The belt-handline of tubular webbing that is described in the section on caving equipment, with the spare carabiner, can be used here. It is a good argument for adopting the belt-handline as a standard piece of caving equipment. If you have neither the belt-handline nor a spare carabiner, try to thread your leather belt around a rung; be sure to rebuckle it before you untie your belay. This is really not much better than no attachment at all, because the belt buckle is not likely to have much strength and the leather will be weak if it is old or cracked. In truth, this procedure should only be considered as a last resort.

Some climbers experience bruising on the inside of their thighs after a long climb on the ladder. This is caused by the protuberance

Figure 61
Climbing a cable ladder against sloping rock.

of the rungs on the outside of the cables, and can be avoided by sewing a 12 in. x 8 in. double strip of denim or other heavy fabric on the thigh area of your caving pants. The problem usually disappears as the climber becomes more experienced.

Rigging and Derigging the Cable Ladder

Overcoming rigging problems may require a certain amount of inventiveness, but there are a number of don'ts that should be kept in mind. Some correct riggings are shown in Figure 62 and some incorrect

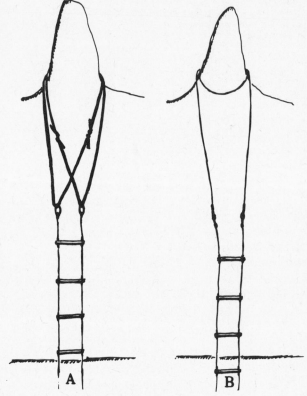

Figure 62
Two acceptable cable ladder riggings.

A. Ladder is suspended by locking carabiners from two separate slings of tubular nylon.

B. Cable rigging with its C links attached to C links on the ladder.

riggings are shown in Figure 63. The site for the rigging should be chosen carefully. The anchor point must be strong enough to support the climber by the criteria outlined in the section on rigging a belay. Climbing a cable ladder in a waterfall and climbing a cable ladder that is resting on a sloping rock are both unpleasant experiences. If you can find an alternative site that doesn't include these miseries, use it.

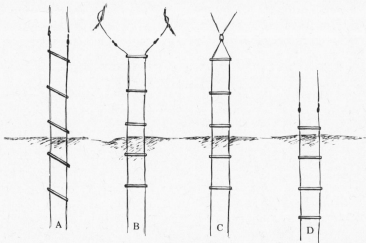

Figure 63
Some incorrect cable ladder riggings.

A. The sloping rungs are diffi-cult to climb.

B. The rigging is spread too wide at the top rung. The sharp angles formed in the wire create two points of possible weakness.

C. The rigging should have two points of support, not just one carabiner.

D. There are too few rungs above the lip of the drop. Climber needs 4 or 5 rungs above the lip to hold onto as he goes over.

Unroll the ladder at the top of the drop and lower one end carefully. Don't tie off one end and then throw the whole roll of ladder over the edge. This is inadvisable for several reasons, not the least of which is that it causes an unholy mess of tangled spaghetti for the first climber to unravel.

Snagging on projections becomes a truly annoying complication after all the party have ascended the ladder and it has to be hauled up before it can be re-rolled. A large protruding snag may make this task impossible. As standard practice, a lightweight pull-back cord (parachute cord) should be attached to the bottom rung of the ladder and the other end should be tied onto something at the top of the drop before the last person goes down the pit. The last caver to climb up the ladder should unsnag the ladder as he climbs while, at the same time, the caver at the top is pulling up the ladder behind him with the cord.

Rolling the Cable Ladder

The method of rolling the ladder as shown in Figure 64 produces a compact (7-inch diameter) coil that is not easily snagged. It requires two people and the ladder must be rolled tightly.

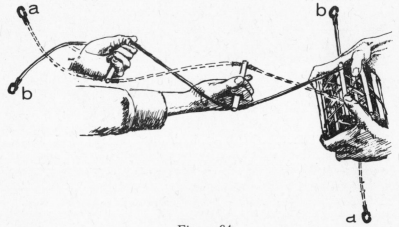

Figure 64
Coiling a cable ladder.
Twist the ladder after every rung. Alternate left-hand twist with right-hand twist. When ladder is coiled, join C link *a* at the end of the ladder to C link *a* in the center of the coil, and C link *b* to C link *b*.

Raising and Lowering Equipment

If you need to raise or lower more than one bag or piece of equipment at a time, don't tie them in one lump. A clump of bags will readily snag on projections. Instead, space them about two feet apart along the rope and tie them on with either an Overhand or Figure 8 knot.

11.   Emergencies in Caving

If you are a novice and are in a decision-making position when a cave emergency occurs, your situation won't be enviable. Inexperience makes decision-making more difficult and more vulnerable to poor judgment. The only protection that you, as a novice, have is adequate preparation for the particular cave that you are about to visit. This means adequate equipment, reliable light sources and spares, plenty of food, sufficient training and practice. Seek as much information as possible beforehand about the nature of the cave: its exact location, its proneness to flooding, its difficulty in relation to your own level of experience. This will give you some indication of what special equipment you will need and whether it is too far beyond your level of experience. Make a list of all this information plus information of the type suggested below, then deposit it with a reliable person—preferably a caver—in the area that you are visiting. If the caving area is very remote or if you know of no one in that particular caving area whose reliability is unquestionable, leave the list with your room mate, a parent, or some other reliable person where you live. If the keeper of the list is a non-caver, advise him adequately on the procedure to follow if you are overdue on your trip. Your listing of information should be similar to the one below.

Sample Emergency List

Name of the cave to be visited, topographic map coordinates, road route, physical features of the trail from the car to the cave, and physical description of the entrance, *i.e.*, large, small, on a hillside, at base of a large outcrop, and so on.
Is the cave known to flood?
The nature of the cave, i.e., predominantly vertical, horizontal, wet, maze, and so on.

Name of the owner of the cave. Is permission to enter needed?
Is special equipment needed to explore this cave? Specify.
Name of at least one person not in your group who knows the cave.
Does a map of the cave exist? Who has one?
Telephone number for the nearest cave rescue team.
Names and telephone numbers of several cavers who could be contacted in case of an emergency.
Expected time of entrance . . . and date.
Expected route underground.
Expected time of exit from the cave . . . and date.
Names of people in your group.
Leader.

Quite needless to say, *all* this information may not be available to you. However, the more of these details you are able to supply, the better will be your chances of survival in an emergency. After gathering this information, you should begin a permanent file on each of the caves you have visited so that future lists can be quickly made if you revisit the cave. If your group has a current mimeographed membership list, this too should be left with the person who is holding the list.
NOTE: You owe the person with whom you deposit the list, the courtesy of a phone call as soon as possible after you exit from the cave. It would be extremely disconcerting to arrange a search for a missing party only to find that the missing party had exited from the cave at the expected time and were home in bed while the search was going on. Twelve hours after the expected exit time is a reasonable time to wait before the keeper of the list should worry, unless unexpected rains have occurred. Those underground will, of course, not know that it is raining outside. A small party should be sent into the cave to warn them (if this is feasible and if the particular cave they are exploring is known to flood).

Obtaining Help in an Emergency

To speed up the operation, the help of the local state police unit should be sought. They will be able quickly and efficiently to locate members of the nearest cave rescue team, or at least local cavers if no rescue team is available. The assistance of cavers should be sought in preference to non-cavers. Local non-caver rescue teams generally come with coverage from the local TV station and newspaper. This is undesirable for many reasons. Such coverage is often a source of misinformation, and if a rescue or search is prolonged, the incident may create considerable local interest and gather a large number of spectators. Onlookers will hinder the operation and cave-owners generally react poorly to strangers in mass swarming over their land. Even a highly skilled fire department or Civil Defense rescue team may be of little use underground if they have no caving experience. An experienced cave rescue team is generally well worth waiting several hours for. Getting an injured caver out of a cave without compounding his injuries is a formidable task. A team with the right equipment for the job and cave rescue experience will be able to remove the caver relatively quickly and with the least chance of further injury. Whether to wait can only be determined by the nature and urgency of the caver's injuries, the extent, complexity, and depth of the cave, and the amount of equipment other than a litter—i.e., ropes, climbing hardware, and winch—that will be needed. Rescues are most difficult in caves with small, serpentine passages and/or water traps. Pit rescues are a lesser problem, but still are best handled either by a pit rescue team or at least by experienced vertical cavers. If no cave rescue team is available, a Civil Defense rescue team may be of invaluable assistance working in conjunction with an experienced group of cavers. The first-aid supplies, stretchers, winches, ropes, and other special equipment usually held by a Civil Defense rescue team may otherwise be unavailable to the cavers.

Getting Lost in a Cave

Sometimes people do get lost in a cave, although this is unlikely if good route-finding technique has been observed. While it is not possible to outline a course of action that can be applied to *all* lost parties of cavers in *all* types of caves, some of the following suggestions will be helpful.

If your party has not left an information list with someone outside the cave and if no one has been told where you are going, the chances that someone outside the cave will know where to look, even if aware of the fact that your return is overdue, are almost negligible. Therefore, you will have to get yourself out of the cave. It is likely that a systematic search in the cave will yield an exit. Keep calm and conserve food and drinking water. Keep the party together. Light can be conserved by using one light for every two people. The caver with the light will have to light the way for the second caver—with all due caution. Follow every passage to its termination, then systematically check out each side passage. A small cairn of stones used as a marker at the beginning of each of these side passages after they have been checked out will reduce the possibility of the same passage being covered twice. If by the time your combined sources of light are almost exhausted, you are still not out of the cave, there are still two alternatives from which you can choose. The first alternative is to turn out all remaining lights, keep each other warm by body contact, carefully dole out food and water, and wait to die. The second alternative, when no one in the party has any light left, is to feel your way around the cave in total darkness until you happen upon the entrance or break your neck, or die of starvation—whichever comes first.

There is a message in this horror story: Leave the information list every time you go caving or at the very least tell someone where you are going. When in the cave, observe route-finding techniques and don't get lost, and leave something at the cave entrance in a conspicuous place. In a very remote caving area, in addition to the above, leave relevant information on a paper tucked under the windshield wiper of your car plus a brightly colored piece of clothing or similar item near the cave entrance where it can be spotted from the air.

For the group which is lost, but which can count on a search party because the information list was left with a reliable person, the prognosis is relatively cheerful. On discovering that the group is lost, the leader should compare the present time with the time that it would take a search party to get to the cave after the alarm had been raised, and within that estimated time period the group should make further intensive effort to get themselves out. If the entrance still hasn't been discovered by the time you think the rescue team is on its way, sit down, conserve light, and keep each other warm until the rescue party arrives. Periodic bursts of shouting will help the rescuers to locate you. There is nothing more frustrating to a search party than to be looking for a group that is constantly on the move. However, a last suggested burst of intensive effort on the part of the lost group to locate the entrance is advisable because it keeps the morale of the group high and the movement will keep people warm. There is also the probability that last burst of effort will be the one that locates the entrance. Find out if a search party is out looking for you before you leave for home after you do find your way out. If so, every effort should be made to reach the searchers and call off the search.

Hypothermia

This is the term used to describe the condition in which there is a loss of heat in the body core, as opposed to body surface temperature. Its symptoms range from violent shivering in its mildest form, through muscular rigidity and lack of coordination, to mental confusion, total apathy, and death. While even its milder manifestations are uncommon in U.S. caves because of their being in the temperate zone it is wise to be aware of the situations in which it is most likely to occur.

Any prolonged immersion in cave water without a wet suit can bring on hypothermia. Standing around in wet clothing in a cave, or sitting still in a cave for any length of time, even when bone dry, can bring it on. Being well prepared for a particular cave means anticipating the likelihood of these hypothermia-producing situations, and taking the right precautions. For example, use a wet suit in a cave where prolonged immersion is possible. If you don't own a wet suit, carry a dry sweater in an ammo box or 2-lb. coffee can with airtight plastic lid, and change out of your wet clothing after immersion before you continue exploration. That spare sweater should be available and added to the clothing you are wearing if you anticipate a long wait in one spot even if you have not been in water.

Reversing the symptoms of hypothermia is possible if it is handled promptly and the victim is removed from the cave as quickly as possible. The first step is to remove the victim's wet clothing and replace it with any dry clothing that can be spared by the rest of the group. If the caver is too incapacitated by hypothermia to be removed from the cave or if he can't be removed for other reasons, he should be kept off the cave floor by seating him on a bed of ropes, cave bags, spare clothing, or anything available. Body heat provided by the rest of the group is effective in facilitating a rise in core temperature. Warm drinks are helpful. It may be necessary, if the victim is severely affected, to carry him from the cave and take him straight to a hospital. In this case, it may be wise to send two cavers out ahead of the victim to call an ambulance. Having an ambulance waiting at or near the entrance will materially aid the victim's chances of recovery.

It is most important that a caver who begins to feel the symptoms of hypothermia should tell the rest of his party. The sooner the condition is detected, the easier it is to reverse.

Panic

This may occasionally be observed in a caver. Calm conversation, advice, and the encouragement, not derision, of the rest of the party will usually defeat it. The frightened person should be encouraged (or even bullied) into taking the necessary steps to extract himself from his predicament or to wait calmly until another person can help. Wherever possible he should be encouraged to help himself, although such encouragement may fall on deaf ears if the victim has already reached the state of panic rigidity. This rigidity is difficult to handle. Explaining to the victim the length of time he will be in his situation before a rescue party can be fetched to rescue him, may help to mobilize a rigid person into looking for ways to help himself. If self-help is not possible, quietly send out of the cave for enough manpower and equipment to rescue him *if* they are not available in the group. Expend considerable effort in reassuring the victim, meanwhile doing all you can to make him more comfortable and his situation safer.

Flooding

Never go underground in overcast or rainy weather unless the cave you are planning to visit has no history of flooding. You can recognize a cave that floods by the twigs, mud, and leaves that are deposited on the walls of the cave at the high-water line. A cave with a wet or dry stream-bed near or in the entrance should be most suspect. Even steady rain can sometimes turn a normally easy and pleasant cave into a treacherous waterway.

If you are trapped by rising water, calmly hurry to the highest place you know of in the cave. Then sit and wait for the water level to drop. A big, high cavern is the best choice.

Leadership in an Emergency

Although each member of the group should be consulted for ideas if a crisis occurs, one person should be responsible for decision-making. This would logically be the most experienced member of the group. This member is, of course, generally the leader of the party. If he be unable or unwilling to accept the responsibility, then the person who feels the most capable of accepting the responsibility should make the decisions.

12. Accidents Involving Injury

Caving is a high-risk activity and, like other high-risk activities, the possibility of accidental injury is always present. Cavers should never be caught completely unprepared. They should have a thorough knowledge of first aid. Uninformed first aid can hasten the death of an injured person, so this chapter is by no means to be considered a substitute for a formal course in first aid. It is supplemental material, dealing only with the most common caving injuries.

In the event of a caving accident, everyone should keep calm. If the victim is bleeding, or not breathing, or has lapsed into a state of shock, first aid will have to be carried out immediately. It may also be necessary to remove the victim to a safe place if he is in danger of further injury. In no case, of course, should the victim be moved until a careful examination has been made and the nature and extent of his injuries has been ascertained. When the immediate problems are under control, take note of the victim and base your treatment on your interpretation of the following details:

1. *Victim's general condition*

Conscious and alert   Disoriented about the events and place
Conscious and dazed   Shivering
Unconscious   Pain—where?
Agitated   Is pain severe or mild?
Lethargic and uninterested

2. *Bleeding*

From the mouth (differentiate between bleeding from a wound in the mouth or bleeding from an unseen source)
From a wound—where and what type of wound?
Was it or is it light or heavy bleeding?
Did the victim lose much blood before bleeding was brought under control?

Was it or is it venous blood (dark and oozing) or arterial blood (bright
red and possibly spurting)?
Possible internal bleeding?

3. *Respiration* (can be determined by placing a hand on the victim's
chest or by watching the rise and fall of his chest)
         Normal   Gasping   Shallow
Was it necessary to restore breathing?

4. *Fractures*
Where?
Simple
Compound (broken ends of bone cutting through the skin, crushed
bone, overlapping ends)

5. *Pulse* (can be felt by placing the forefinger over blood vessel on inside of the wrist)
         Slow   Irregular   Racing   Normal

6. *Skin*
Clammy   Hot   Pale
Cold   Normal   Sweating

7. *Eyes*
Both pupils dilated   One pupil dilated   Normal

The results of your examination should make it possible to diagnose
and treat some of the more obvious injuries. These results should
be recorded and taken out with the cavers who have been sent for
rescuers.

Under the subheadings below, symptoms of the injuries most commonly encountered in caving accidents and the recommended first-aid
treatment will be discussed.

Wounds and Bleeding

It can take as little as one minute to bleed to death, so act quickly
to control bleeding. Most bleeding can be controlled by raising the
injured part, then applying direct finger pressure to the wound. Before
pressure is applied, however, a pad of clean cloth should be placed
over the wound. If blood saturates the pad, apply more layers—don't

remove the original pad. When bleeding has stopped, leave the dressing in place and secure it with tape or a bandage. If the bleeding is not controlled by this method, maintain pressure on the wound and instruct an assistant to press on the nearest pressure point. These points can be found in the places where a large artery passes near the skin surface and they are located on the inside wrist (pulse), in the groin, and at the inside upper arm, and in the neck near the jaw. Never use this jaw artery to control bleeding. Any interruption to the blood supply to the brain can cause permanent brain damage. *Note:* Do not use a tourniquet unless every other control has failed.

Shallow, graze wounds should be washed with clean water, if available, and covered with a clean dressing. However, cave water is usually dirty and therefore unsuitable. If no other water is available, it is better to apply a thin layer of antiseptic cream and a dressing until the wound can be properly cleansed later with clean water.

The possibility of internal bleeding should not be overlooked in a serious caving accident. The symptoms are similar to those of shock and there may also be bleeding from the mouth and nose if the wound is in the lungs, stomach, or within the skull. In such cases, hospitalization is urgently needed. A second kind of internal bleeding occurs with bone fractures and may be serious. In a complicated fracture, many blood vessels can be severed. Bleeding into the surrounding tissues will occur and may be sufficiently heavy to precipitate shock. Splint and elevate the fractured part. Purple or red discoloration may be apparent around the bone if internal bleeding is present.

If the wound is not in the abdomen or chest and the patient is conscious, small drinks can be given at intervals to maintain the body fluid balance and allay thirst.

How to Restore Breathing

Mouth-to-mouth resuscitation is the fastest and best method to restore breathing in a drowning accident or other accident that results in the cessation of breathing. A person who has stopped breathing will die in 4 to 6 minutes, so waste no time in beginning resuscitation.

1. Place victim flat on his back.
2. Turn the victim's head to one side. Remove gum, food or other loose matter from his mouth with your fingers.
3. Tilt his head back as far as possible and keep it back during resuscitation.

4. Pinch his nostrils shut and with your mouth tightly over his, blow in hard about every 5 seconds.

5. Each time after you blow in, put your ear near his mouth and listen for air to rush back out. His chest should expand when you blow in and contract as the air comes back out. He is breathing when his chest is moving without help.

6. When breathing is restored, replace any wet clothing with dry clothing and keep him warm.

Shock

While shock is predominantly triggered by physical trauma, the psychological factors that can alleviate or worsen shock should not be overlooked. One of the many advantages of having a first-aid expert at the scene of an accident is that there will be at least one person who is seen by the victim to be doing something to save him. This helps to allay the victim's fear and increases his positive feelings about his survival. It is easy to imagine the victim's increased terror while lying bleeding and in pain, when he realizes that not one of the onlookers knows what to do. This terror may influence the severity of the shock and speed the onset of shock symptoms. Even when the injury is not particularly serious, panic can produce symptoms that equal in severity those brought on by serious injury. In short, the *appearance* of efficiency is very important to the victim because it provides him with very necessary reassurance.

Shock is a depressed condition of many of the body functions following injury. It will almost certainly accompany heavy bleeding, severe pain or complicated fractures. Inexpert handling of the injured part will produce or worsen the shock. If the victim remains in a shocked state, death will result, even though the injury may not otherwise be fatal. Therefore, the prevention and proper handling of shock is extremely important. *Treatment for shock should be a routine part of first aid.*

The symptoms are general weakness, skin that is clammy, pale and cooler, and the breathing may be shallow and irregular. The victim may be mentally detached from the situation and he may be nauseous. Faintness should not be confused with shock, even though the symptoms are similar.

To prevent and treat shock, make the patient lie down. Make a bed for him of spare clothing, ropes and anything else that will keep him off the cold floor. Replace any wet clothing and surround him with any warm clothing that can be spared by the rest of the group.

Body contact with other cavers will provide warmth and psychological comfort. Raise the head and chest a few inches if breathing difficulties are present. Raise the lower part of the body if blood loss has been great, unless there is suspected head injury or if breathing becomes more difficult as a result. Warm, sweetened fluids can be given in small increments to help maintain body warmth and relieve thirst unless the patient is unconscious, nauseated, or has suspected internal injuries. Don't give alcohol to accident victims.

Unconsciousness and Head Injury

Unconsciousness in an accident is generally either caused by breathing failure or by head or spinal injury. Do not apply mouth-to-mouth resuscitation as previously described unless breathing has stopped. If breathing is normal, treat unconsciousness as head or spinal injury. If fainting is recurring, suspect head or spinal injury.

Diagnosis of head injury should be based on the history of the accident and the presence of swelling or a wound on the head. Pupils may be unequal in size and bleeding in the ear, nose or mouth is sometimes present. There may be paralysis of one or more of the extremities. Headache, dizziness and disorientation are common symptoms. When any factors, even doubtful ones, point to head injury, treat it as head injury. Keep the victim warm, still and comfortable, and do not attempt to move him out of the cave, even if he is conscious and willing, until a stretcher evacuation is possible. Do not give him drugs and give nothing by mouth.

Whenever a person has pain in the back or neck following an accident, treat it as a spinal injury. This is a serious injury and if it is inexpertly handled, it can permanently maim or kill the victim. Keep him absolutely still and as flat as possible. Do not move him unless his life is threatened by his immediate surroundings. Evacuate him from the cave only when a medical expert, preferably a doctor, is present to supervise.

Fractures

The two most common types of fractures are simple fractures (closed) and compound fractures (broken bone ends cutting through the skin, overlapping ends, crushed bone, broken ends grinding against each other).

Symptoms of a fracture are any or all of the following: *(1)* tenderness in the body part containing the fracture or tenderness at the probable site of the fracture, (2) swelling, (3) abnormal shape. The victim will not want to move the injured part. He may be able to move it (although he shouldn't), but he will usually prefer to hold it quite still to avoid pain. However, pain is not always present. Treat *suspected* fractures, dislocations, and sprains as fractures. Many "sprains" turn out to be fractures when X rays are seen.

First-aid treatment begins with immobilizing the injured part before moving the victim. This consists of making sure that broken ends and nearby joints cannot move, bend, or twist. The simplest splints and those most immediately available in a cave are self-splints. Self-splinting is no more than immobilizing the injured part by tying it against a normal part, *i.e.*, broken leg against other leg, broken arm against the body. However, regular splints such as commercial wooden or inflatable splints, rolled newspapers, straight-tree branches or other rigid objects are more effective (if applied correctly) in preventing further injury than are self-splints. Regular splints should be applied before the victim is moved out of the cave unless the fracture is minor and not painful. Self-splints are most useful when you need to move the victim out of further danger in the cave, or when you move him to a more comfortable place. A splint should be padded before application and needs to reach above and below the break. It should not be placed directly over the break. Tie it on with handkerchiefs, strips of cloth, or proper slings.

The pelvic fracture victim should be placed on his back and transported from the cave in this position. The victim may have no pain while lying still, but will probably complain of pain in the crotch or lower abdomen when he moves.

Rib fractures need no immediate treatment unless breathing discomfort is experienced. It is difficult to diagnose rib fracture unless the victim has tenderness at certain points along the rib. Rib fractures are serious if the broken ends are driven inward where they could puncture the lungs or other soft tissue. If the victim is not in great pain and the rib fracture appears to be a simple one, he may be able, with the help of his companions, to get himself out of the cave.

Moving the Victim

When first aid has been applied and the victim is as comfortable as possible under the circumstances, *how* and *when* to move him out of the cave will have to be decided upon.

First, consider moving the victim out of the cave with the help of the cavers who were with him at the time of the accident. If this can be done safely (for everyone) without the victim having to use the injured part, and if he is willing to try, then this will usually be the best thing to do. The cave is cold and uncomfortable and exposes the victim and his companions to the added danger of hypothermia. Therefore, a self-help rescue will reduce the chance that this complication will occur.

In many cases, however, either the type of injury or the type of cave will make full-scale rescue necessary. If the wait is going to be a long one, someone should go out of the cave to fetch down sleeping bags, spare clothing, fuel for warming food and drinks, and spare carbide and water. These items will make the enforced sojourn more comfortable and safer for all concerned.

When formal rescue is necessary, write down all pertinent information about the situation to send out with the cavers who go for help. Record all information about the injuries and general condition of the victim—based on the check-off list that was suggested previously —so that suitable medication and equipment can be brought back. Detailed information about the type and number of cave obstacles that will be encountered by the rescuers when they evacuate the victim will facilitate the proper selection of rescue equipment. Do everything possible to get a doctor or at least, a first-aid expert back to the victim if the injury is serious. Apart from these situations requiring a major rescue effort, there will be many incidents in which a knowledge of first-aid will stand you in good stead.

Reading List

British Caving: An Introduction to Speleology, by the Cave Research Group, edited by Cecil Cullingford. Boston: Routledge and Kegan Paul.

The Caves Beyond, by Joe Lawrence Jr. and Roger Brucker. New York: Funk and Wagnalls, 1955.

Caves of Adventure, by Haroun Tazieff, 222 pp. New York: Viking Press, 1953.

Celebrated American Caves, edited by Charles E. Mohr and Howard N. Sloane, 339 pp. New Brunswick, N.J.: Rutgers University Press, 1955.

Depths of the Earth, by William R. Halliday, 398 pp. New York: Harper and Row, 1966.

Descent of Pierre Saint Martin, by Norbert Castaret. New York: Philosophical Library, 1956.

First Aid Textbook, rev. ed., edited by American Red Cross. New York and Garden City: Doubleday & Co., Inc., 1957.

The Life of the Cave, by Charles E. Mohr and T. Poulson. New York: McGraw Hill, 1966.

Manual of Caving Technique, edited by Cecil Cullingford, 416 pp. Boston: Routledge and Kegan Paul, 1969.

Mountaineering: The Freedom of the Hills, edited by Harvey Manning, 2nd edition, 430 pages. Seattle: The Mountaineers, 1960.

Potholing: Beneath the Northern Pennines, by David Heap. Boston: Routledge and Kegan Paul, 1964.

Speleology: The Study of Caves, by W. George and Brother G. Nicholas, 120 pp. Boston: D.C. Heath, 1964.

Ten Years Under the Earth, by Norbert Castaret, 240 pp. London: J. M. Dent and Sons Ltd, 1939.

NATIONAL SPELEOLOGICAL SOCIETY PUBLICATIONS
(Available from the N.S.S. Office)

Prusiking, by Robert Thrun. This will be one chapter of the forth-coming *Cavers Handbook* by members of the N.S.S. To date, this is the only source of detailed information on ascending a fixed rope and it is highly recommended to all pit cavers.

N.S.S. Information Series
  1. Information for Neophyte Cavers, 2 pages.
  2. Techniques of Cave Mapping, 2 pages.
  3. The Hard Hat, 2 pages.
  4. Cave Clothing, 2 pages.
  5. Public Relations and Caving, 3 pages.
  6. A Microvelocity Anamo Meter, 4 pages.
  7. Belay Technique, 5 pages.
  8. Methods to Find New Caves, 2 pages.
  9. Belay Tie-off, 2 pages.
 10. At the Other End of the Rope, 1 page.
 11. Scaling Poles.
 12. Belay Rules, 1 page.
 13. Cable Ladders, 5 pages.
 14. Building a Wet Suit, 5 pages.
 15. Bats, 2 pages.
 16. Map Symbols, 2 pages.
 17. Cave Rescue, 8 pages.
 18. Use of the Cave Radio in Mapping, 7 pages.
 19. A Magnetic Induction Cave Radio, 5 pages.
 20. A Nutritionally Adequate Caver's Ration, 8 pages.
 21. A Caving Camera, 1 page.
 22. Selection of a Caving Camera, 7 pages.

1968 Speleo Digest, a collection of the most significant articles on caves, technique, equipment, science, and humor published by caving organizations in 1968.

American Caving Accidents, Issues are available for 1967, 1968, 1969.

N.S.S. News, a caving magazine issued monthly to N.S.S. members.

N.S.S. Bulletin, a scientific publication covering cave-related topics issued quarterly to N.S.S. members.

The reading list given above represents some of the most significant publications that have appeared in recent years. Inquiries about other titles should be directed to the N.S.S. Librarian. Those references which have the most value for the topics around which this book is constructed are as follows:

CABLE LADDER CONSTRUCTION
Manual of Caving Technique, pp 57–108.
"Cable Ladders," N.S.S. Information Series No. 13.

FIRST AID
First Aid Textbook.
Manual of Caving Technique, pp 284–300.
Mountaineering: The Freedom of the Hills, pp 298–312.

CAVE RESCUE AND SAFETY
"Cave Rescue," N.S.S. Information Series No. 17.
British Caving, pp 536–544.
Manual of Caving Technique, pp 325–359.
"American Caving Accidents," all issues.

ROCK CLIMBING AND BELAYING
Mountaineering: The Freedom of the Hills.

CAVE SURVEYING
"Techniques of Cave Mapping," N.S.S. Information Series No. 2.
"Map Symbols," N.S.S. Information Series No. 16.
British Caving, 509–534.

CAVE SCIENCES
Speleology: The Study of Caves.
The Life of the Cave.
British Caving.

INDEX

Food, 27, 28–29
Fractures, 117–118

Group, size of, 46–47

Half hitch, 43
Hard hats, 23, 24–25
Hypothermia, 111

Indirect belays, 62
Injuries, 113–119
    see also First aid

Kneepads, 30–31
Knots, 41–44

Laid rope, 34–35
Leader and leadership, 45–46, 112

Mechanical ascenders, 93
Moving through a cave, 49–52

National Speleological Society, 18, 45, 75

Overhand knot, 43

Pack, *see* Cave pack
Panic, 111–112
Prusik knots, 44, 76, 82, 84, 85
    left-hand, 87
    materials for, 88–90
    right-hand, 77
Prusiking:
    chest harness for, 88, 89, 90
    difficulties, 84–87
    negotiating overhangs, 86–87
    predicaments in, 91
    slings and equipment for, 91, 93, 94–95
    technique, 76–83

Pulleys, 34
    in belay rigs, 56
    in practice rigs, 63, 91, 92

Rappel:
    body, 72, 74
    carabiner wrap, 75
    hasty, 74
    rack, 68, 75
    with tandem brake bar rig, 68–69
Rappelling:
    cautions, 73–75
    definition of, 68
    equipment for, 69
    technique, 69–73
Rescue, 108–110, 119
Rigs and rigging:
    for belaying, 55, 56
    brake bar, 68–69
    for cable ladders, 104–105
Rock-climbing, 52
    belaying for, 64
Ropes:
    description of, 34–36
    maintenance and care of, 37–38

Shock, 116–117
Signals, 64–66
Speleothems, 15, 16
Squeezes, 49–50
    safety in, 50–52

Tandem brake bar rig, 68–69

Vertical techniques, 96

Water, 27, 28
Webbing, 36
Wounds, 114–115